Offshore Engineering

Editor
T.V. Ramakrishnan

GENE TECH

2014

Gene-Tech Books
New Delhi - 110 002

ISBN10 81-89729-80-2
ISBN13 978-81-89729-80-6
EAN: 9788189729806

Published by **: GENE-TECH BOOKS**
4762-63/23, Ansari Road, Darya Ganj,
NEW DELHI - 110 002
Phone: 41562849
e-mail: genetechbooks@yahoo.co.in

Laser Typesetting **: Classic Computer Services**
Delhi - 110 035

Printed at **: Chawla Offset Printers**
Delhi - 110 052

PRINTED IN INDIA

Preface

Today many number of countries are giving prime importance to the subject Offshore Engineering. Most nations that have major offshore oil activities in their economic zones have established regulatory agencies to control and supervise their development.

Most structures in harbours, on coasts in estuaries and deep rivers are carried out by public bodies.

The oceans present a unique set of environmental conditions which dominate the methods, equipment support and procedures to be employed in offshore construction.

The present book explains about offshore platforms and their construction, machinery drives need, and also about number of topics which are related to offshore engineering.

I shall regard my efforts amply rewarded if the book is received with the enthusiasm.

I wish to express my sincere thanks to Gene-Tech Books for their continued interest in bringing out this edition.

T.V. Ramakrishnan

Preface

Contents

Chapter 1

Introduction

Many offshore projects relate to the development of an ocean resource, especially oil and gas. Therefore the sequence and relationships of the parties involved will help constructors in planning their operations.

For offshore oil and gas development, leases for specific offshore areas are granted by the sovereign nation having jurisdiction to a petroleum company, with provision for payments, royalties, taxes, and conduct of operations. The latter may involve specific agreements regarding alliances with other contractors, training and employment of nationals during the construction period, use of local contractors, fabricators, and suppliers, purchase of local materials, areas in which work will be carried out, and research and educational activities to be supported. Many of these act as constraints upon the construction contractor, who may be required to associate with a local partner or a national enterprise. Constructors may be subject to restrictions regarding the number of foreign workers, including skilled workers and supervision, that they may employ. Often they are required to build new construction equipment in that country or, if they bring in outside equipment, to do so under bond assuring its subsequent export. Federal U.S. laws–for example, the Jones Act–prohibit the use of foreign-registered dredges within U.S. waters and restrict the use of foreign towboats.

Following lease arrangements, the oil company will carry out extensive geophysical investigations, including seismic surveys. At this time it may also get shallow core borings, bathymetric data, and environmental information. Exploratory drilling is usually carried out by floating vessels, drill ships and semi-submersibles being the most used in deep offshore areas and jack-ups in more limited depths. In the Arctic and in shallow waters, exploratory drilling may be carried out from bottom-founded mobile structures. These vessels are collectively called "mobile offshore drilling vessels."

The petroleum company, having carried out geophysical exploration and exploratory drilling, may confirm a discovery. It now carries out delineation drilling to determine reservoir characteristics. It also intensifies planning for offshore structures and development, carrying out feasibility studies and preliminary engineering to select a concept and contractors.

Most often, these studies are carried out by one or more engineering contractors or integrated engineering-construction firms. However, some of these may be carried out by in-house teams of the petroleum company with the aid of consultants. At the same time, the company proceeds to obtain more-accurate, site-specific geotechnical and environmental data for use in design. Arrangements must be made for shore-based facilities. Environmental impact statements must be prepared. Financing arrangements are made.

During this entire period, the oil company operator will often have put together a consortium of companies to participate in the project. These may include the national oil company and from 1 to 20 other petroleum companies. Usually, but not always, the operator has the largest percentage, except perhaps for the national oil company. The position of operator is much sought after by petroleum companies, as it confers control of the project, usually with substantial fees to cover management and overhead costs. It also enables the company to develop advanced in-house engineering expertise and management capabilities.

Offshore construction, with its many current opportunities and its tremendous demands and challenges, is emerging as one of the most exciting fields of engineering practice, one which will test human-kind's ability to rise to new heights of skill and courage.

The range of offshore structures is very great.

Most structures in harbors, on coasts, in estuaries, and deep rivers are carried out by public bodies. The dominant current practices in the United Sates are to select an engineer to prepare plans and specifications or, alternatively, to perform this work with an in-house engineering staff. Then, separate construction contracts are awarded, usually on the basis of competitive bidding. A growing trend worldwide is a single contract for design and construction. In the case very large projects, such as major overwater bridges, the contract may include financing. Both BOT (build, own and transfer) or BOOT (build, own, operate and transfer) contract are employed. These contracts include the engineering design to meet the basic criteria established by the agency. Constructibility is thus a major consideration by the contending constructors.

Most nations that have major offshore oil activities in their economic zones have established regulatory agencies to control and supervise their development. These governmental agencies are typically assigned responsibility for ensuring safety during development and operation with respect to the following:

1. Prevention of pollution
2. Prevention of loss or waste of the resource
3. Prevention of injury and death to personnel working on or in conjunction with the development.

National agencies (*e.g.*, the Minerals Management Service of the U.S. government) have established rules for the conduct of offshore minerals development, including especially offshore oil and gas. These rules provide for a review of design, fabrication, and installation by an independent, third-party "verification agent." The constructor is intimately involved in the fabrication and installation phases and must therefore not only submit properly prepared and documented plans for approval, but must carry out the work in compliance with these documents and sound construction practice. Provision must also be made for quality assurance during construction. This may or may not be part of the verification process but, of course, must be integrated with it.

Following completion of the platform, development drilling is carried out from the platform. In many cases, production will start after a few wells are completed, with concurrent drilling continuing. After all wells are operating, work overs must be carried out from time to time in order to ensure the continuing productivity of the wells. Water and gas injection may be instituted to enhance well productivity. Subsea satellite wells may be tied into the primary platform. On the platform, the gas and oil must be separated. Produced water and sand must be separated out, with the water requiring treatment prior to discharge into the sea. In some cases, oil is stored at the offshore platform, usually by the saltwater displacement system. Discharge ballast water must have the hydrocarbons removed by separators to the concentration permitted by regulation.

Most shipment of oil and of gas is by pipeline to a shore terminal. Some gas is used on board to power the platform operation. Flaring of gas from the platform is usually limited by regulation to the initial period of operations and to emergencies. Oil may also be shipped by tanker. In this case, a submarine pipeline from the platform runs to a loading buoy for direct transfer by means of a swivel head to the tanker.

During the operating life of the platform, maintenance must be carried out, repairs performed, and modifications made. While these are usually relatively small contracts in magnitude, they frequently are equally demanding in terms of technical skill and specialized equipment.

When the field has reached the end of its economic lifetime, usually after 20 to 30 years, regulations of most countries provide that all facilities be removed to several meters below the mudline. The wells are capped and cemented off, and the top side equipment is removed. Piles and well casing are cut off. The platform is now dismantled and removed.

For other types of marine construction, whether it be in river harbors or coasts, an Environmental Impact Statement is required and a permit must be obtained by the operator. The conditions attached to this permit are binding upon the constructor and have the force of law. Similarly, the Occupational Safety and Health Act requires full compliance by the constructor.

The oceans present a unique set of environmental conditions which dominate the methods, equipment, support, and procedures to be employed in construction offshore. This same unique environment also, of course, dominates the design. Many books have addressed the extreme environmental events and adverse exposures as they affect design. Unfortunately, relatively little attention have been given in published texts to the environment's influence on construction. Since the design of offshore structures is based to a substantial degree upon the ability to construct, there is an obvious need to understand and adapt to environmental aspects as they affect construction. These considerations are even more dominant in many coastal projects where breaking waves and high surf make normal construction practices impossible. To a lesser extent, they have an important role in harbor and river construction.

Currents, even when small in magnitude, have a significant effect on construction operations. They obviously have an influence on the movement of vessels and floating structures and on moorings. They change the characteristics of waves. They exert horizontal pressures against structural surfaces and, due to the Bernoulli effect, develop uplift or downloading forces on horizontal surfaces. Currents create eddy patterns around structures, which may lead to scour and erosion of the soils. Currents may cause vortex shedding on piles, tethers, and piping.

Even before the start of construction, currents may have created scour channels and areas of deposition, thus creating surficial discontinuities at the construction site. The vertical profile of currents is conventionally shown as decreasing with depth as a parabolic function. Recent studies in the ocean and on actual deepwater projects indicate however, that in many cases, the steady-state current velocities just above the seafloor are almost as high as those nearer the surface. There are substantial currents in the deep sea, just above the seafloor.

There are several different types of currents: oceanic circulation, geostrophic, tidal wind-driven and density currents, as well as currents due to river discharge. Currents in a river vary laterally and with depth. The higher river currents usually occur near the outer edge of a bend. River currents are also augmented locally around the head of jenties and groims. Some of these may be superimposed upon each other, often in different directions.

The worldwide ocean circulatory system produces such currents as the Gulf Stream, with a relatively well-defined "channel" and direction and velocity of flow. Other major current systems exist but are often more diffuse, having a general trend but without the characteristics of a river. Thus the prevailing south easterly trending current along the California and Oregon coasts gives an overall southward movement to sedimentary materials from river outflows.

These major currents may occasionally, often periodically, spin off eddies and branches; the lateral boundaries of the current are thus quite variable. Strong currents may thus occur many miles from the normal path of a current such as the Gulf Stream. Within local coastal configurations, a branch of the main current may sweep in toward shore or even eddy back close to shore.

Recent research has indicated that many of these current "streams" are fed by upwelling or downward movements of the waters and that there are substantial vertical components. These will become of importance as structures are planned and built in deeper waters and will require that accurate measurements be taken at all depths, for both vertical and horizontal components of the current.

Another major source of currents is tidal changes. The stronger tidal currents are usually in proximity to shore but may extend a considerable distance offshore where they are channeled by subsurface reefs or bathymetry. While they generally follow the tidal cycle, the frequently lag it by ½ to 1 hour; thus a tidal current may continue flooding on the surface for a short period after the tide has started to fall.

Actually tidal currents are often stratified vertically, so that the lower waters may be flowing in while the upper waters are flowing out. This is particularly noticeable where tidal currents are combined with river currents or where relatively fresh water of lower density overlies heavier saltwater. This stratification and directional opposition also occurs at the entrance to major bodies of water, *e.g.*, at the Strait of Gibraltar where evaporation from the Mediterranean produces a net inflow.

Since tidal currents are generally changing four times a day, it follows that their velocity and direction are constantly changing. Since the velocity head or pressure acting on a structure varies as the square of this current velocity, it can have a major effect on the mooring and control of structures during critical phases of installation. The current velocities are also superimposed on the orbital particle velocities of the waves, with the pressure and hence forces being proportional to the square of the vectorial addition.

While in regular harbor channels the tidal currents may move in and out along a single path, at most offshore sites the shoreline and subsurface configurations cause the directions to alter significantly, perhaps even rotate, during the tidal cycle. Ebb currents may be directed not only 180° but often 150°, 120°, or even 90° from flood currents, and this variance itself may change periodically. Tidal currents may reach speeds of 7 knots (3.6 m/s) and more.

River currents, especially those of great rivers with large discharges, such as the Orinoco, extend far out to sea. Because the density of the water is less, and perhaps because of silt content, the masses of water tend to persist without mixing for a long period; thus substantial surface currents may reach to considerable distances from shore. River currents may, as indicated earlier, combine with tidal currents to produce much higher velocities on ebb and reduced velocities on flood.

Wind persisting for a long period of time causes a movement of the surface water that is particularly pronounced adjacent to shallow coasts. This may augment, modify, or reverse coastal currents due to other causes.

Deep-water waves create oscillatory currents on the seafloor, so that there is little net translation of soil particles due to waves alone. When, however, a wave current is superimposed upon a steady-state current, the sediment transport is noticeably increased, since its magnitude varies as the cube of the instantaneous current velocity. The vertical pressure differentials from the waves lift the soil particles, which are then transported by the current.

Adjacent to the shore, the translational movement of the waves produced definite currents, with water flowing in on top and out either underneath or in channels. Thus a typical pattern of the sea will be to build up an offshore bar, over which the waves move shoreward and break on the beach. This piles excess water on the beach, which may move laterally, then run out to the sea. The out flowing current cuts channels in the offshore bar. The seaward-flowing current becomes the infamous "undertow." These lateral and seaward-flowing currents may be a hazard or may be taken advantage of to keep a dredged channel clear through the surf zone.

In the deeper ocean, currents are generated by internal waves, by geostrophic forces, and by deeply promulgated eddies from major ocean streams such as the Gulf Stream. It appears that currents of magnitudes up to 0.5 knots exist on the continental shelf and slope and that currents up to 2.6 knots (1.3 m/s) can be found in the deep ocean.

Strong currents can cause vortex shedding on risers and piles, and vibration of wire lines and pipelines. Vortex shedding can result in scour in shallow water. It can also result in cyclic dynamic oscillations of cables, tethers, moorings, and vertical tubulars, such as piling, which can lead to fatigue. Vortices occur above a critical velocity, typically 2 to 3 knots. These vortices spin off in a regular pattern, creating alternating zones of low pressure. Vortices and whirlpools can form at the edge of obstructions to river flow, *e.g.*, around the end of groins or at the edge of an underwater and wave, leading to severe local scour.

Chapter 2

The Offshore Platforms

Introduction

Offshore platforms of the gravity-base category are designed to be founded at or just below the seafloor, transferring their loads to the soil by means of shallow footings. Such gravity base platforms have usually been constructed or reinforced and prestressed concrete but a few have been built of steel or a hybrid of concrete and steel.

These platforms are almost always constructed in their vertical (final) attitude, enabling much or all of the deck girders and equipment to be installed at an inshore site and transported with the substructure to the installation site. These structures are usually self-floating, although, when necessary, additional lift forces may be developed by temporary buoyancy tanks or special lifting vessels.

To minimize soil-bearing loads, these structures have a large base "footprint." To provide buoyancy, they have large enclosed volumes. They thus generate much greater inertial forces under waves and earthquake, 50,000 to 100,000 tons of lateral force being typical, with special structures developing even more. Thus sliding tends to become the dominant mode of failure, at least for water depths up to 150 to 200 m.

To transfer this lateral load into the soil and thus prevent sliding, steel skirts and steel dowels are employed, designed to penetrate and thus force the failure surface farther below the seafloor. Such skirts also provide protection against scour and piping. While the skirts are typically fixed to the base of the platform during fabrication, in special cases where shallow water limits draft, skirts or spuds may be installed through sleeves after the structure has been seated on the seafloor .

The construction of a typical concrete gravity-base platform takes place in a well-defined sequence of stages. For each stage, there are several important criteria which must be met:

1. The structure must be watertight and have stability and freeboard at all stages of construction.
2. The loading conditions and combinations acting on the structure are significantly different from one stage to the next. Structural integrity must be assured at each stage.
3. Ballasting and compressed-air systems (if these latter are employed) must be carefully and positively controlled at all stages.

To meet the above criteria, it becomes necessary to control weights and dimensions with great care. These structures are very large and massive, extending 100 to 200 m or more on the three axes.

Most errors to date have been due either to overlooking an intermediate stage or to combining two or more stages to save computational effort. Detailed sketches of each sub-stage, along with evaluation of the pertinent hydrostatic, hydrodynamic, and structural loadings, must be prepared to enable visualization by both design and construction engineers.

The internal subdivisions of the structure are subjected to differential pressures, primarily due to the different ballast water heads acting on each side. Compressed air may be used on occasion to pressurize a compartment; the pressures occasioned thereby must be considered.

Accidental conditions must also be considered: the loss of compressed air from under the base skirts on one side, rupture and flooding of one compartment due to collision from a boat, a broken ballast pipe, or a failed penetration. Under these accidents, the structure may be permitted to suffer minor local distress so long as its integrity, stability, and buoyancy are maintained. Progressive collapse–for example, where one compartment floods, overloading the adjoining bulkhead which in turn fails, and so on–cannot be permitted.

A more-detailed description of the special requirements and considerations at each step and stage will now be presented.

We can understand the elevation of stratfjord B platforms in Figure 2.1.

Construction Stages of Platforms

During the construction in basin side slopes must be protected against excessive erosion under heavy rain, and more importantly against slope failure.

Access is frequently underestimated. At least two well-surfaced roads should lead into the basin or else trestle access should be provided at the top. Similarly, the roads around the base raft on the floor of the construction basin should be well-drained and properly

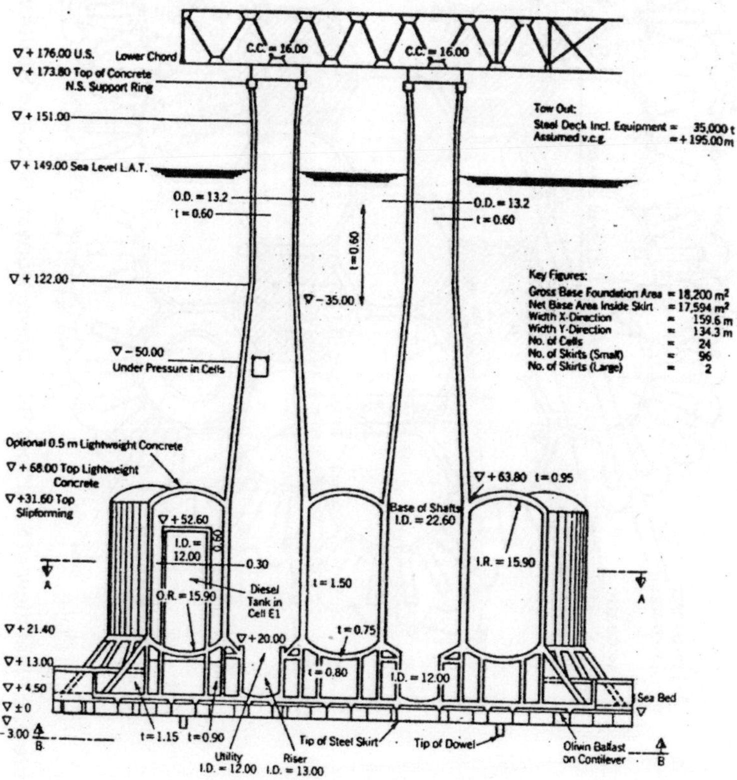

Figure 2.1: Elevation of Stratfjord B Platform

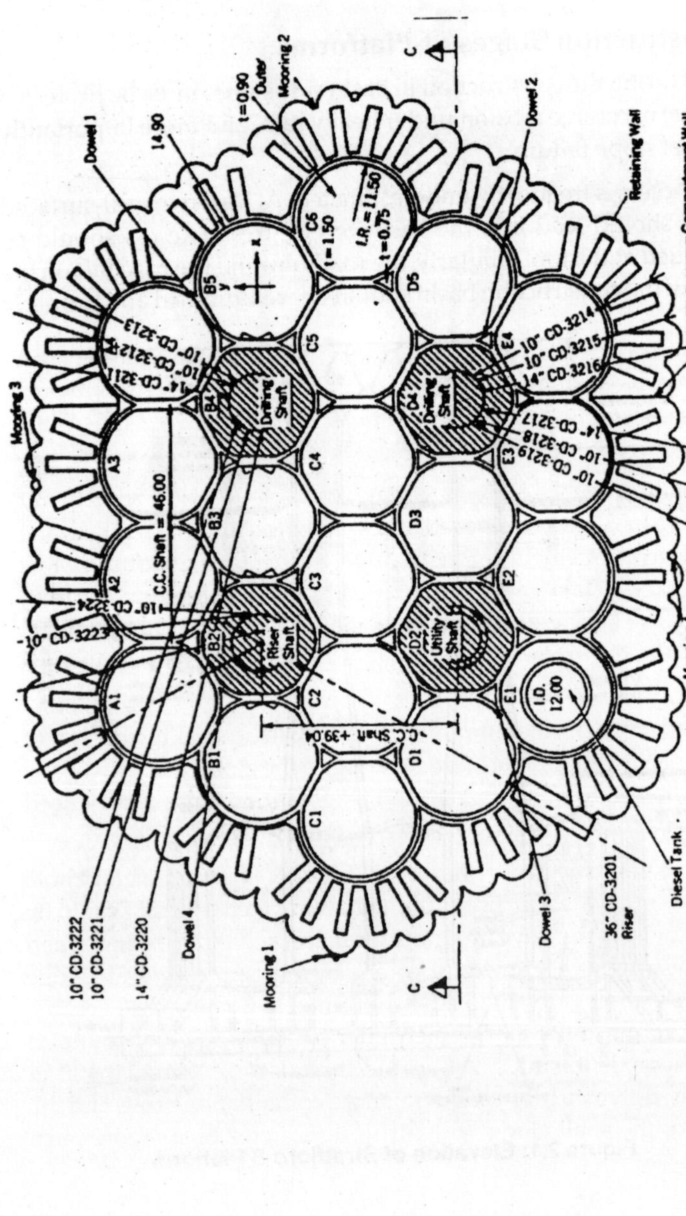

Figure 2.2: Plant View of Stratfjord B Platform, Norway, Showing Struts which support Cantilevered Base Slab.

surfaced, *e.g.*, gravel, since a basin tends to collect large quantities of water and mud.

Caisson gates are constructed where several uses of the basin are planned; they can be removed and replaced in 1 to 2 days. Such gates were installed after the piping incident in Queensland, and they are installed at the Highland Fabricators basin in Nigg Bay and Howard-Doris's basin at Loch Kishorn, both in Scotland. Two details need to be verified: first, the provision of drainage from under the gate caisson so that no uplift can occur; second, the connections

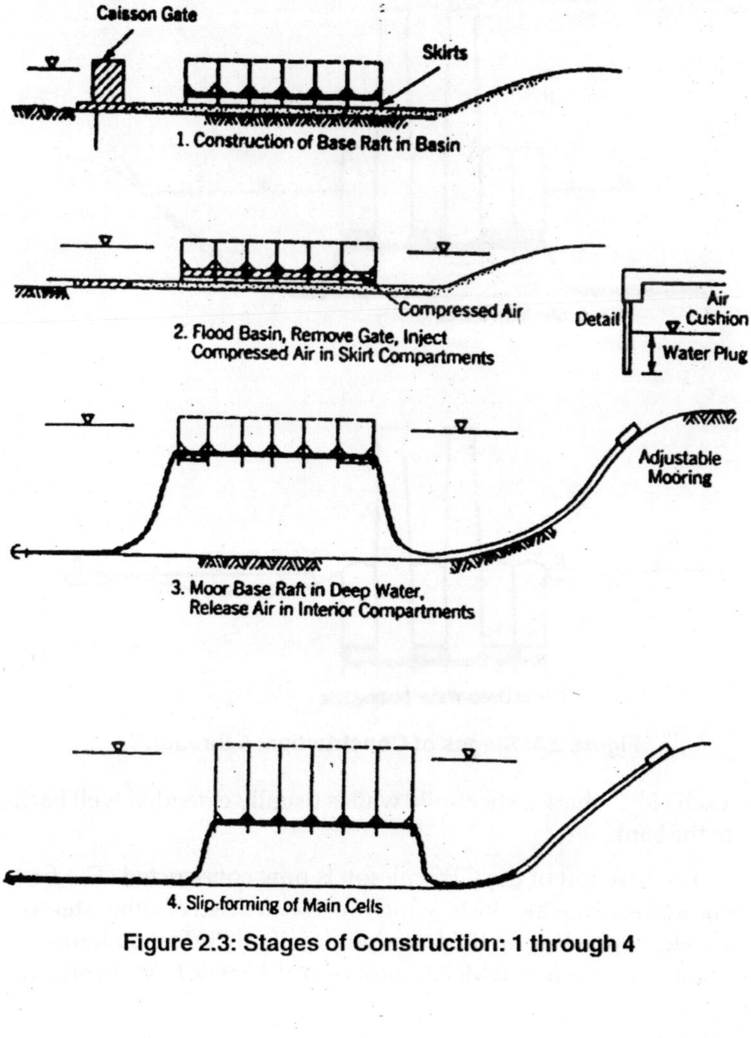

Figure 2.3: Stages of Construction: 1 through 4

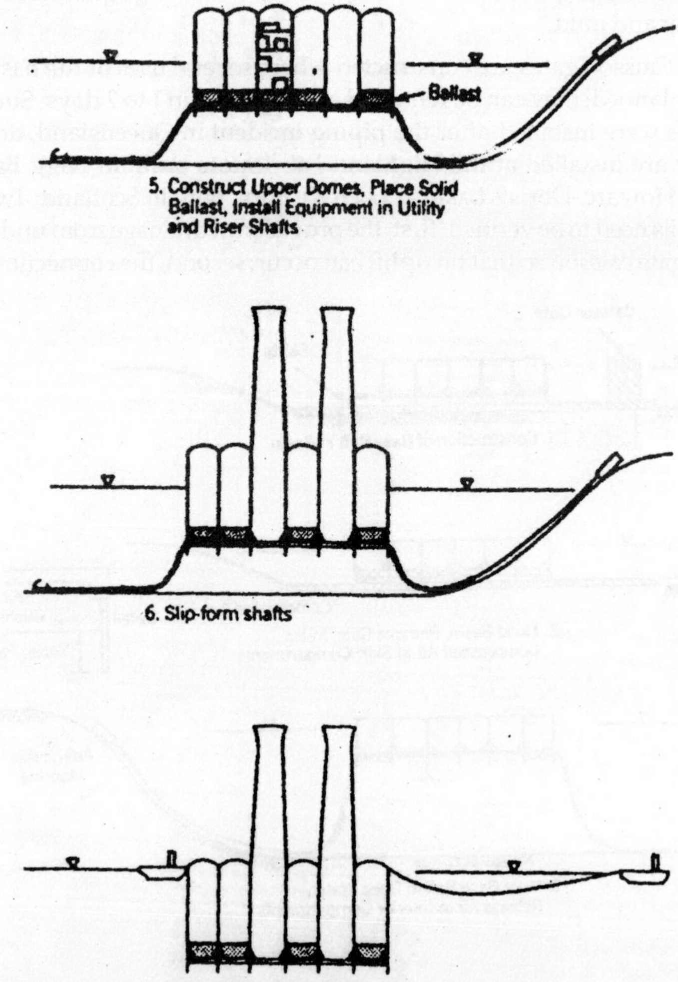

Figure 2.4: Stages of Construction: 5 through 7

at each side, where a sheet pile wall is usually extended well back into the bank.

The base raft of the GBS caisson is now constructed. The first items installed are the skirts, which are constructed of either steel or concrete, depending on the foundation soils at the installation site. In some cases, such as at McAlpine's yard at Ardyne Point, Scotland,

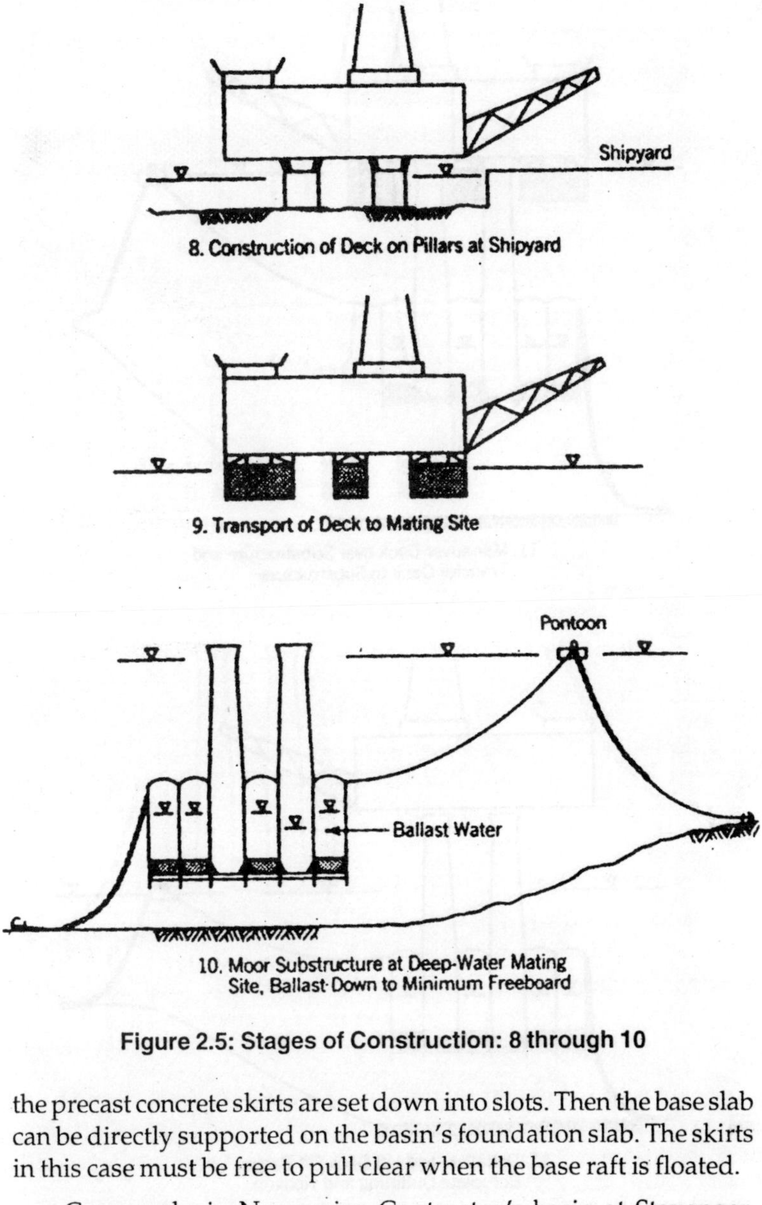

Figure 2.5: Stages of Construction: 8 through 10

the precast concrete skirts are set down into slots. Then the base slab can be directly supported on the basin's foundation slab. The skirts in this case must be free to pull clear when the base raft is floated.

Conversely, in Norwegian Contractor's basin at Stavanger, where longer steel skins are employed, the skirts are supported on the basin's foundation slab and the structure's base slab is

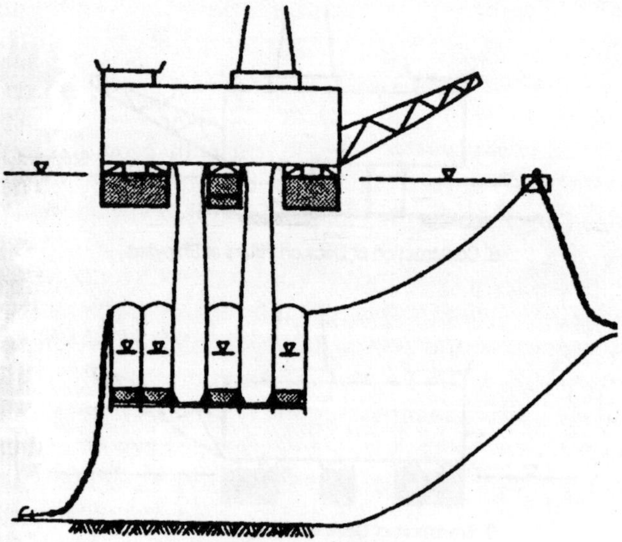

11. Maneuver Deck over Substructure and Transfer Deck to Substructure

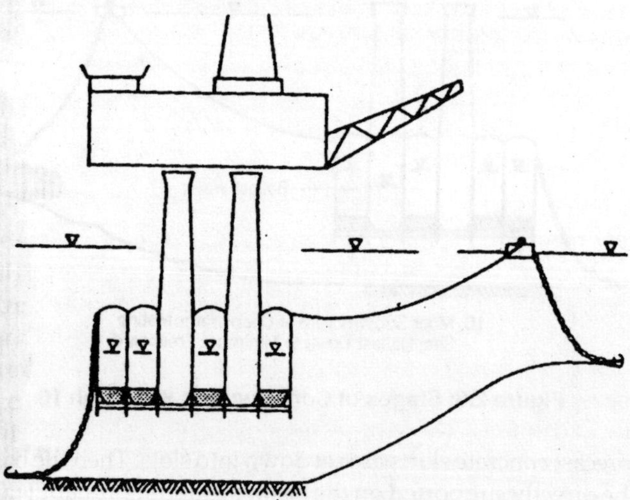

12. Deballast and Lift Deck Off Barges. Complete Outfitting and Hookup

Figure 2.6: Stages of Construction: 11 and 12

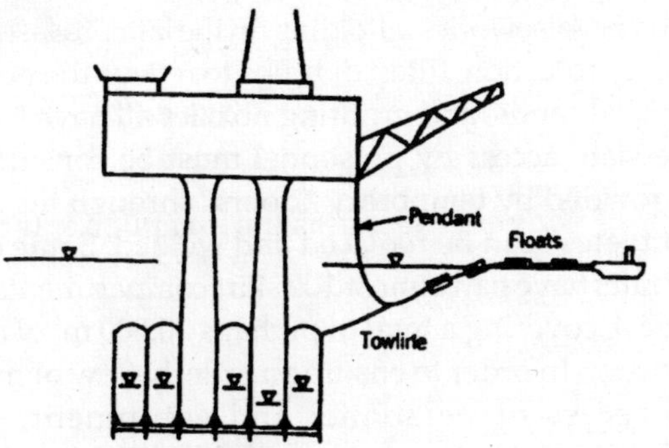

13. Tow to Field

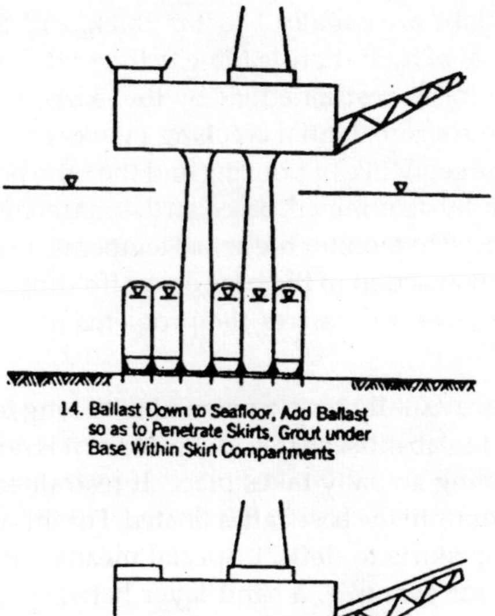

14. Ballast Down to Seafloor, Add Ballast so as to Penetrate Skirts, Grout under Base Within Skirt Compartments

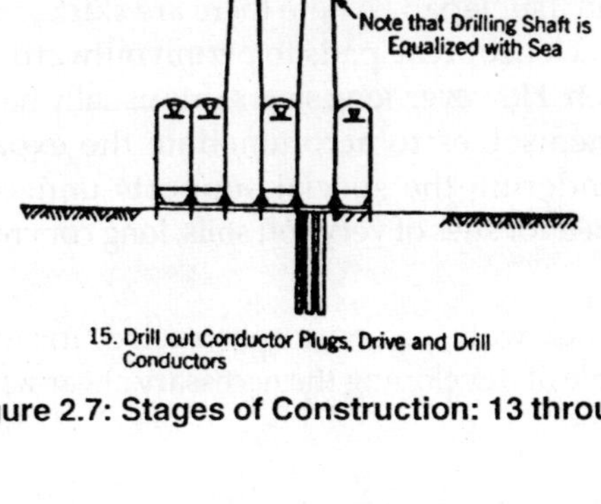

15. Drill out Conductor Plugs, Drive and Drill Conductors

Figure 2.7: Stages of Construction: 13 through 15

constructed on falsework scaffolding. In the latter case, where anodes for cathodic protection, filtered drains to relieve the pore pressures in the surficial sands, and grouting nozzles all have to be installed under the slab, access by personnel must be considered. This is usually provided by temporary "doors" through the skirts. Later, the "door plates" can be replaced and welded. Some of the larger GBS base rafts have had almost 100 skirt compartments, all more or less identical, covering a total of perhaps 20,000 m^2 of area and 500 m of perimeter. In order to ensure an orderly flow of materials and efficient access of personnel and equipment, these skirt compartments must be clearly marked; otherwise a major portion of any worker's time may be spent hunting for the right location. Concrete base slabs are usually 1 to 2 m thick, and they typically generate high heat of hydration, leading to thermal expansion. If the subsequent cooling is restrained, as by the skirts, scaffolding, or adjacent slab pours, substantial cracking may occur. To minimize this problem, the sequence of pouring and the time between pours needs to be carefully determined, based on thermal analyses. Thermal probes may be used to monitor the actual temperature in the interior of a slab. The construction of the base on scaffolding, raised above the floor of the dock, minimizes the problems of constraint and cracking.

Base slabs are usually post-tensioned with long tendons; once again the concrete slab must be free to shorten as it is not prestressed until the shortening actually takes place. If restrained by friction, this will not occur until the base raft is floated. For this reason, when there are no long skirts to deflect, special means are provided to reduce friction–for example, a sand layer between polyethylene sheets, on which a plywood soffit is laid. This was the method used during the construction of the CIDS concrete structure at NKK's shipyard dock in Tsu, Japan. Where there are skirts, these may be supported on thick neoprene pads to permit outward and inward shear deformation. However, long steel skirts usually have adequate flexibility in themselves to accommodate the expansion and contraction, rendering the special supports unnecessary. For platforms destined for sites of very soft soils, long concrete skirts are constructed.

The lower raft walls are now constructed, forming cellular partitions, capable of developing the necessary shear when the base raft is floated.

If there is an upper slab on the base raft as an integral part of the design configuration, it will provide the deck of a barge like structure. Then any hog-sag moments during floating will be resisted by the top and base slab. Unfortunately, having such a top slab on the base raft is the exception rather than the rule. Usually there is only the multitude of intersecting cell walls. There is no upper flange, no "deck.," for moment resistance during float-out. The base raft is then a large "barge" with minimal depth and strength.

Unequal moments arise, for example, because of a thickened outer wall or because of cantilevered extensions to the base slab which will be submerged during float out and/or mooring. Because the center of gravity of a semicircular annulus lies outside that of a complete semicircle, a significant hogging moment may be induced. This may be aggravated by the weight of the skirts, since these are usually concentrated around the outside, and by cranes, mooring chains, and the vertical component of mooring forces.

Reduction of the hogging moment may be attained by increasing the air cushion under the outer skirts and by adding ballast to the center cells. However, this latter usually conflicts with the requirements to take all possible steps to reduce draft during float-out. A more positive solution, often adopted, is to prestress the top of the cell walls and/or add reinforcing steel to offset or resist the hogging moment and reduce residual stresses. Note that since many of the larger GBS base rafts are not symmetrical, it is usually necessary to check bending around a number of possible axes, both orthogonal and oblique.

With the base raft structure nearing completion, the mechanical systems are installed, principally consisting of the saltwater ballast piping, under base grouting system, skirt drainage and skirt venting, instrumentation such as strain gauges on skirts, bottom clearance acoustic sensors, and base mat pressure cells.

A careful inspection is now made of all skirt compartments to remove the accumulated debris, forms, and scaffolding that inevitably are left. The doors in the skirts are now welded closed. Ballast water is pumped into the cells to hold the raft on its foundation slab.

With the structure ready for float-out, final checks are made of weight and displacement to ensure that flotation can be accomplished within the design tolerances of draft and heel. Note that for the typical 120-m-diameter raft, it takes only 1° heel to increase draft by 1 m. A minimum underkeel clearance of 0.5 m is usually required. If a tidal cycle is involved, the clearance must extend over a sufficient period to accommodate potential delays in the float-out operation. The basin is now flooded.

To reduce draft, air cushions are usually pressurized under the skirts. Compressed air may be introduced through the under base grouting system, for example. To prevent the air from escaping under the tip of the skirts, a "water plug" about 0.75 m high is usually left. If the skirts, for example, are 4 m long, then the use of the air cushion can reduce the draft by 3 m or so, depending on the area covered by skirts which are capable of holding an air cushion.

Now let us study about float-out stage and about mooring. When the flooding is complete, the air cushions are tested for each set of skirt cells to be sure there are no leaks. The ballast water is removed, and the base raft floats up to its float-out draft. All systems are checked.

Meanwhile the exit channel has been re-sounded to ensure adequate draft, and dragged or profiled (with an electronic-acoustic profiler) to ensure against any rocks, debris, piles, or other obstructions above grade. If all is in order, the gates are now removed. If the "gates" consist of sheet piles and a rock dike, this area must also be carefully swept to ensure against obstructions.

Navigational aids will have been established to guide the exit and to mark the route to the deep-water construction site. Laser or light ranges are useful in monitoring the effect of transverse currents. Electronic distance-measuring and position-plotting systems are usually installed on a temporary control deck on the base raft, supplemented by a sextant or theodolite.

Moorings will have been set at deep-water site, secure to mooring buoys. All boat traffic in the area will have been stopped. Weather reports will have been double-checked, especially with regard to wind. Tidal predictions will have been verified for the specific exit site.

Float-out from the basin has to be very carefully controlled because, as noted earlier, the typical raft is very unstable and weak in bending and hence sensitive to accidental loads and events. Winches mounted on the walls or sides of the basin control lateral movement, while highly maneuverable tugs pull the base raft out into the channel. The intact freeboard will be a minimum of 1 m above the local wave crest height with allowance for run-up. Because mooring lines can jam in a sheave or otherwise be fouled, contingent means are available to cut the lines if necessary.

The base raft is now towed to the deep-water site. A harbor crane barge will aid in securing the mooring chains from the mooring buoys to the base raft.

The air cushion will still be contained within the skirts. The release of this air must follow a carefully calculated procedure to ensure against overstressing the base raft in bending. In general, the air cushion under the central portion can be released initially, while that under the periphery cells is kept until additional freeboard is attained by the extension of the peripheral wills. Alternatively, a short cofferdam can be built on the outer walls.

Control of the air pressure under skirt cells is very tricky. If pressure gauges are used in different segments of the base, the readings may lead inherently to the wrong action. When the Statfjord A base was floated out, pressure gauges were the only instrumentation used. As the base raft tilted slightly, the pressure on the downside (B) rose, and that on the upside (A) fell. So the natural reaction was to bleed air from the downside (B) and add air to the upside (A). This caused increased heel, continuing to show higher pressures on the downside and lower pressures on the upside. So, in effect, the "corrective" action was slowly jacking the platform to ever greater heel.

Earlier reference was made to the low stability of the base raft. This is largely due to the air cushion during float-out which creates a free surface in each skirt compartment, reducing the gross water plane moment of inertia by the sum of the individual water planes. Any ballast water in an internal compartment will also shift to the downside, adding to the overturning moment.

The proper means of control, therefore, is not pressure but volume. Water level indicator gauges–for example, based on electrical resistivity or hinged floats should be installed in the skirt compartments to give a measurement of the height of the water plug relative to the base slab and hence a measure of the air volume. Base raft level indicators of suitable sensitivity should also be installed at the control station.

During float-out, mooring, and initial construction operations at the deep-water site, there is always the possibility for loss of air pressure under one set of skirt compartments. The subdivisions must therefore be carefully selected to minimize the adverse effects of loss of air pressure or of local flooding.

With deep concrete skirts and a deep air cushion, such as those for the Draugen platform, the air pressure at the top of a compartment will be substantially higher than the outside water pressure at that

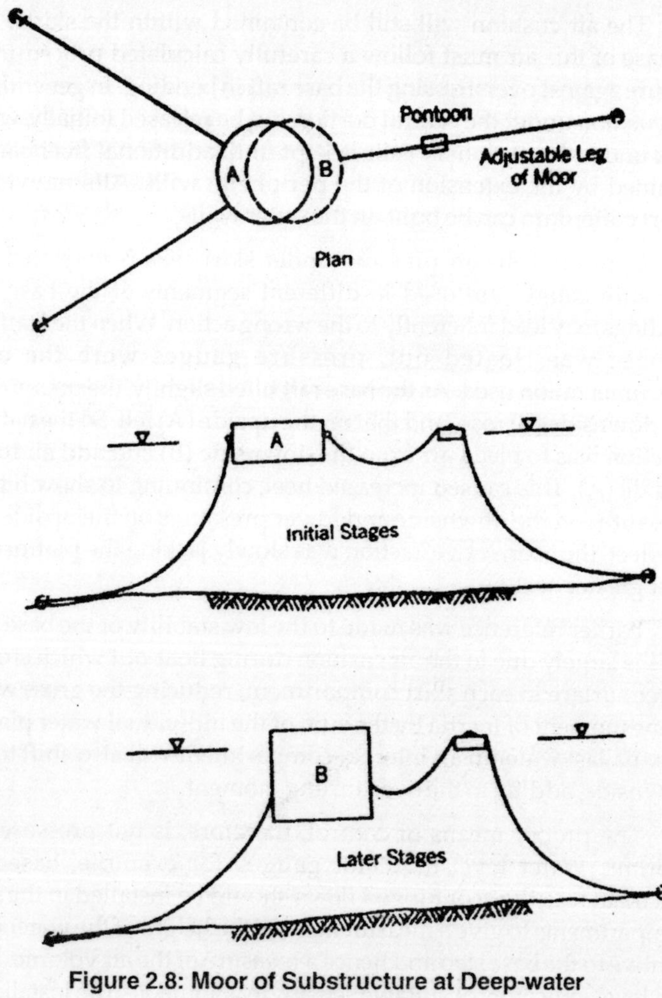

Figure 2.8: Moor of Substructure at Deep-water Construction Site

level. The structure must be checked for this resultant internal overpressure, which creates tension in the cells.

Either three or four mooring legs will have been pre-installed, terminating in mooring buoys arranged around the GBS deep-water construction site. These will typically consist of a heavy clump anchor on the seafloor or drilled-in anchor on the beach, several shots of chain, and then a wire rope mooring line or chain to the buoy. The buoy is held in approximate location by a separate small clump anchor and pendant.

From the buoy, a chain is run to the base raft and connected by shackles. Now there is a continuous mooring from base raft to anchor, with a float supporting it in the center, giving it some spring under load and reducing the tilting of the base raft. At least one line will have provision for progressive take-up as the construction proceeds and the mooring sinks lower in the water. In addition to moorings, a water line and a power line are usually run out from shore. By mating these lines to a wire rope, adequate strength can be provided.

The principal construction activities for the platform now take place while the structure is moored in adequately deep but protected water. During this period, the structure must be safely protected against accidental flooding. Detailed consideration must be given to the sequences of construction since temporarily unbalanced loads will result in structural stresses which may be locked in as residual stresses.

Shortly after the base raft of the Ninian central platform was moored, a storm arose. Continuous wave action beating against temporary bulkhead closures near the waterline caused working and then fatigue of the bolts holding the closures on. The external compartments on that side flooded. The water pressure inside that compartment ruptured the temporary internal closures to two adjoining compartments. Fortunately, all other closures held and the raft did not heel further.

However, the concrete batching and mixing barge had been moored tightly to the far side. Its manholes on deck were not secured. When the base raft listed into the waves, the far side of the base raft rose, lifting the adjacent side of the concrete bare, consequently submerging its deck on its far side. The water poured into the manholes, and the concrete barge broke loose, turned over, and sank. On all temporary closures, therefore, the bolts should be designed to resist fatigue. Double nuts with lock (spring) washers can be used. Large bearing plates can be installed to prevent local crushing of timber struts.

A procedure which can be usefully employed to increase the freeboard during the critical stage of float-out and initial caisson construction is to extend the outside walls higher than those on the inside. On the Ekofisk and Hibernia caissons, for example, the outside walls were kept built up above the inner concrete. This can also be accomplished by extending a steel cofferdam up from the base raft, cantilevered from the base raft to give adequate freeboard.

During construction afloat, the platform is required to maintain its 1 m freeboard under the 100 year storm wave plus run-up, for that location and season of the year, as well as one-compartment stability, that is, to be safe with any one compartment flooded. If this cannot be maintained, then the intact freeboard must be increased to about 6 m.

Construction can now take place at the deep-water site. If the GBS is being constructed of reinforced concrete, as has been most common to date, then transport and installation of reinforcing steel, prestressing ducts, and concrete must take place in substantial volume and on a more or less continuous basis.

Where labor supply is limited, where there is an unusually large concentration reinforcement, or where embedments must be set with great accuracy, then slip forms may not be applicable. Panel forms, cantilevered up, or flying forms are then employed, with each pour being defined as to extent and duration. This often necessitates the use of window boxes and tremies ("elephant trunks") to place the concrete. It also means that there will be a large number of horizontal construction joints. Both these requirements can and have been successfully met, not only in base raft construction for offshore platforms, but also in the rather similar walls of nuclear reactor containment structures.

Slip forms have also been employed in such cases, subdividing the structure into segments, with vertical construction joints. The Ninian central platform, with its seven concentric walls, was so subdivided. Precast concrete elements have been extensively incorporated into base raft construction, the Ninian central platform and Global Marine's Super CIDS being prime examples. These should be cast in as large segments as practicable; 200-ton segments were used on the Ninian central platform. They can be incorporated into the remainder of the structure to act in monolithic fashion by means of dowel extensions and cast-in-place concrete joints or by post-tensioning.

Match-cast jointing techniques appear applicable in the case of many precast assemblies; these can follow the technology now perfected for long-span bridges, where units up to 25 m wide have been successfully joined by this method. Match-casting was employed with great success for the breakwater wall of the Ninian central platform, where horizontal joints were employed. Match-casting not only led to rapid construction but also assured perfect alignment of the complex post-tensioning ducts.

Other construction expedients include the use of precast concrete soffit forms to support horizontal slabs and domes during construction of the base caisson. The Seatank platforms used precast concrete pyramidal forms which were domed shells constructed of shotcrete. A number of the Condeep platforms have used precast shells to form the lower dome. The Hibernia platform used precast beams and slabs to form the roof.

Steel soffits with internal support beams built in can be similarly used, the whole designed for composite action with the cast-in-place concrete. These steel soffits can have superimposed trusses which support them and which are later embedded to serve as reinforcement.

When extremely thick walls are required, as in some designs for ice-resistant structures, precast concrete hollow box units may be employed, to be filled later with cast-in-place concrete. Examples of the effective use of combined precast concrete and cast-in-place concrete are the outer walls of the Ekofisk caisson and the Ninian central platform.

With steel GBS structures, large prefabricated segments are being lifted, guided into place, aligned, and welded. They must be given temporary support against wind. Welding locations must be protected from rain and spray. Columns and shafts must be adequately stiffened to prevent out-of-roundness distortions. Erection pins and bolts are needed to assure proper fit-up.

For all structures being constructed afloat, access must be provided for a large number of personnel. This matter has often been underestimated in the past. Personnel must move from shore to a barge or floating dock alongside, up and over the outside walls, and across open cells and the internal walls to their site of work. These walls are constantly being extended up, with closely spaced reinforcing bars, ducts and climbing rods (for the slip-form yokes) all projecting above the walls. The freeboard of the structure is also changing. To add to the difficulties, distances between cells and walls are large, extending the reach of cranes to their limits and sometimes necessitating rehandling of reinforcing steel and other materials.

Clear identification of cells and walls is essential so that there will be no confusion by workers about where they are to go. Access routes must be planned and safe walkways provided. Reinforcing steel bars may purposely be spread at certain specific locations; this should be incorporated in the working drawings and checked by the designer rather than being improvised in the field.

The most recent gravity-base platforms have required very dense concentrations of reinforcing steel. Bars have been bundled and spacing between bars reduced. Mechanical connectors have been used to splice bars, rather than laps. With such concentrations of steel, up to 500 kg/m^3, the placement has controlled the rate of progress.

Concrete mixes must be designed that are sufficiently workable to flow around and through these tightly placed bars. Coarse aggregate sizes have been reduced to 8 to 10 mm and the sand content increased to 50%.

Placement of concrete has generally been accomplished by a combination of pumping and bucket or even wheelbarrow delivery, the latter for distribution into the slip forms. Retarding admixtures are added as necessary and adjusted for temperature, to match the rate of rise of the slip forms.

Freeboard can be kept constant by selective ballasting of the cells. However, these operations must be carefully planned and checked at every stage and sub-stage. On the Beryl A platform, main cells were ballasted with water up to 40 m deep. No ballast water was placed in the small star cell interstices between the primary cells. No notice was taken of this because these interstices were covered over by temporary timberwork platforms that rise with the slip forms. Small in plan as they were, they had never been designed nor checked for a hydrostatic head of water ballast inside the main cells, placing tension and shear on the connecting walls. Only after conclusion of concreting of the base caisson to a height of about 52 m, when the work platforms were removed, was the large and extensive cracking in the walls made apparent. The integrity of the structure was finally restored, but only through a mammoth effort of concrete repair.

This illustrates the need in planning to portray each of the many incremental steps of construction, stage by stage, to reflect the step-by-step addition of concrete, which changes weight, trim, and draft, the progressive changes in ballast, change in stability (usually not critical), the changes in hydrostatic loading, and the structural response to the different load combinations.

In particular, the matter of differential hydrostatic head on the internal and external walls must be addressed. In the usual case, the net load is the difference between two large numbers, each the square of the water head. Mere application of a load factor to the differential may be totally inadequate to cover tolerances in these heads. Conversely, application of a load factor to the larger of the two heads of, say, 1.3, along with 0.9 to the lesser head, can result in impracticably, almost ridiculously, high design loads, especially as the total head (depth) increases. Therefore, it is necessary to establish realistic tolerance ranges for the water ballast in the various compartments and the ballast differential should be physically limited by provision for overflow.

During construction afloat, the draft of the base structure gradually increases. This changes the scope of the mooring lines. Usually, the necessary tension can be maintained by taking in on one, or at the most two, of the mooring lines. This is facilitated if one of the lines can be run to shore, assuming it is nearby. If not, adjustment means can be provided by a large mooring pontoon, moored over the bright of the line, taking it up vertically to shorten the effective scope. Such a system is essentially a large spring buoy and may be excessively flexible; this is why a large pontoon is desirable.

Taking up tension in one leg of the moor by hauling in at the structure presents practicable difficulties because of the underwater connection and the large forces involved. However, this has been done on bridge pier caissons, which present a similar problem, by using a cable- or chain-grip hydraulic tensioner, working through a watertight sleeve built into the structure. From time to time, the mooring line has to be stopped off, the sleeve extended upward, and the tensioner relocated to a new center of rotation.

The moorings must be designed to hold the substructure during any storm wind plus waves and currents having a significant probability of occurrence during the construction period. Usually this is selected as the 10-year return storm, but considering the seriousness of the consequence of a rupture of the moorings and the delay and loss to the total project, a longer-return-period storm should perhaps be adopted.

The substructure almost always has several large barges moored to it, and it may be impracticable to move these away prior to a storm that arises suddenly, as was the case at Ninian. Therefore, these should be included in the calculations.

Crew changes are another practical problem due to the large number of workers per shift and the need to carry on work continuously. During rough weather and high winds, there arise the problems of getting personnel on and off the boats, seasickness, and finding position and course in fog, rain, or snowstorm. These must all be addressed in the planning stage.

Provision must also be made for emergencies. What is to be done if a worker is injured? Are there pallets or basket stretchers available which can be handled by a tower crane to hoist the injured person to a boat? What about man overboard? A continuous safety boat has been provided 24 hours a day around some of the larger platforms. Life preservers and trailing floating lines should be provided. Life jackets should be worn by all personnel working over the side and during transport. It is unnecessary and perhaps counterproductive to try to have personnel wear them while working in the cells; they may catch on the reinforcing steel. However, if there is ballast water in the cells, life jackets will be appropriate.

Cells eventually are 60 m or so deep. Safety nets may be needed where personnel are working across the top and to catch falling debris when personnel are working underneath. Usually these nets are provided at least for the principal shafts. Projecting reinforcing bars are a puncture hazard and an eye hazard. They should all be fitted with a red plastic cap. Prestressing ducts, especially vertical ducts, become convenient receptacles for tools, bottles, aggregate, and the remains of lunches. They should all be covered with red plastic caps, which will also serve to keep out the rain. On one project, a heavy rain was followed by freezing weather, the rainwater ran into the ducts and then froze and split the concrete walls in multiple fractures.

Fresh water and electric power must be supplied to the structure during construction afloat. If the deep-water site is relatively near to shore, a hose and power cable may be run, attached to a wire line. Otherwise a water barge and generator barge must be moored alongside.

During the final installation process, many stages later than those being discussed at the present, the platform will be ballasted down onto the seafloor. As it nears touchdown, the water trapped under the caisson tends to cause the platform to skit laterally, more or less uncontrollably. While venting of the skirt compartments and reducing the rate of descent will minimize this problem, a more effective method has been developed to prevent dislocation which uses three or four dowels that engage the seafloor while there is still 2 to 3 m of bottom clearance.

These dowels cannot be installed in the basin because of draft limitations, so they are lowered and fixed in position at the deep-water construction site. A typical dowel is a steel tubular 2 m in diameter, with 75- to 100-mm-thick walls, extending 4 to 5 m below the skirts. In any specific case it is designed to provide adequate lateral resistance to hydrodynamic skidding by developing passive resistance in the soils, which, of course, produces shear and moment in the dowels. The dowel is designed so that failure, if it occurs, will first be by plowing through the soil and then by bending/buckling of the dowel, so that the structure itself will remain undamaged. As soon as the structure is moored at the deep-water site, each dowel is lowered and secured in place by grouting within a sleeve.

During this phase, major mechanical system installations take place, primarily the crude oil piping and oil level indicators. Several complex pumping and ventilating modules may be placed in the utility shaft. In addition, risers for flow lines from satellite wells and for crude oil transfer lines are installed in or on the periphery of the base caisson. The risers may include J-tubes and, in some cases, curved conductors. In other cases, entry tunnels may be constructed, through which the pipelines will eventually be pulled.

The next substage at the deep-water site is the construction of the domed roof of the base caisson. The construction of this domed roof, perhaps 10,000 to 20,000 m² of area, covering over the 15 to 30 cells, often requires more time on the construction schedule than the construction of the walls of that base caisson, even though the ratio of concrete may be as low as 1:3. The delay, of course, is due to support of the forms. Prefabricated dome soffits of concrete have been employed as well as the more conventional steel trusses and plywood. A steel soffit, containing all the necessary reinforcing steel and having multiple studs for composite action, could conceivably be placed as a single lift, permitting rapid concreting thereafter.

Once concreted, the domes are often covered with a meter or so thickness of lightweight concrete to absorb the impact of objects dropped from the deck during drilling and production operations. Over-dome pipe supports are also provided for any crude oil transfer piping which may have been installed on the exterior of the base caisson and which now must be led horizontally into the riser shaft.

Since the substructure may later be towed to another site for deck mating, towing attachments and notches are fitted around the periphery of the walls where they intersect the domes. Towing attachments are designed to withstand the failure load of the towing lines but to have a weak link themselves which fails before pulling out and damaging the structure.

Circumferential prestressing is also usually required at this intersection between domes and exterior walls. Proper staging should be applied to facilitate this operation.

Most offshore platforms require solid ballast in order to lower the center of gravity as much as possible. Solid ballast may consist of concrete placed above the base slap, within the cells. Mass concrete typically generates high heat of hydration and hence leads to thermal expansion which may adversely load the adjoining walls and crack the slab. Therefore, the mix should be selected for low heat and low modulus of elasticity. High density is desirable; this may be enhanced by selection of aggregate of high specific gravity. Blast furnace slag-cement, low-heat cement, or a portland cement-pozzolan mix should be employed. Strength is normally not a criterion; hence total cement content can be kept relatively low. To prevent undesirable structural interaction between the ballast concrete and the walls, sheets of crushable material (polyurethane foam) are often placed on the boundary walls. Where the ballast concrete is also designed to act structurally in composite action, it may be necessary to place it in sections to avoid causing distress in the walls due to heat of hydration expansion.

Iron ore has also been used effectively as solid ballast, placed by pumping in as a slurry. Provision must be made to decant the slurry by drainage. Finally, where solid ballast is to be placed on exposed slabs, such as cantilevered base slabs, high-density rock can be placed. There are several methods available for placement of rock on exposed underwater slabs: placement through a large-diameter (*e.g.*, 1m diameter) tremie tube, placement in a skip or bucket that is lowered to the slab before discharge, and discharge from the surface in relatively small batches. Impact force on the slab has to be considered.

Tests both on offshore platforms and on similar operations have shown that discharge in a small but coherent mass is usually quite satisfactory, without excessive segregation, especially if the mass is flooded before discharge so that it does not contain large quantities of entrapped air. The alternative, applicable in deeper water, is placement through a tube.

The next stage is typically the construction of the shafts, which typically may be from one to four in number. These are usually tapered, with varying wall thickness, necessitating the use of sophisticated adjustable slip forms. Shaft slip forms, like chimney (stack) slip forms, tend to rotate. To correct this after it has started is difficult. Chain jacks may be used to react against the climbing rods, or a more rigid steel beam stub may be embedded in the wall and used as a reaction point.

The large single shaft of the Ninian central platform was constructed using match-cast precast concrete segments, later post-tensioned to act monolithically.

Control of verticality is by lasers, set accurately on the base slab. Unlike chimneys on land, the vertical axis of a floating structure changes with list and trim of the structure, so all points must be relative to the base slab.

During construction of the shafts, vertical access is provided by a construction elevator, which is laterally supported at intervals. Inserts for support must be located in concrete that will be adequately strong by the time the elevator tower needs to be connected.

Post-tensioning of the completed shaft is carried out, most often from a gallery in the base raft. At this stage, the gallery will be deeply submerged and hence subject to very high hydrostatic pressures and differential pressures from the ballast water. Since this is a temporary facility for the purpose of jacking the tendons and grouting the ducts, additional internal supports of structural steel or even timber may be found more practicable than permanent concrete struts.

The grouting, carried out under sufficient pressure to reach to the top of the shafts, as well as the wicking action of the strands, leads to considerable sedimentation and forced bleed. Provision of means for refilling at the top and use of a thixotropic gelling admixture are usually both necessary.

On some gravity-base platforms (*e.g.*, the Andoc Dunlin platform) the upper portion of the shafts was made of structural steel in order to reduce weight and overturning moment. The large "cans" of the shaft were set by use of a large floating derrick after the base structure had been ballasted down to reduce the height above water.

Gravity platforms have also been constructed of steel; examples are the four small Loango platforms off the mouth of the Congo River and the large Maureen platform in the North Sea. Their construction generally follows that of the concrete platforms except that, being light, most of the structure can be constructed in the construction basin.

Large "bottles," cylindrical tanks attached to the edge of the base, provide both buoyancy and stability during tow and installation.

An especially daring hybrid concept has been proposed by Saga Petroleum of Norway. The mating sequence proposed involves joining the top section of steel tubular framing to the lower section of concrete cellular structure while both are inclined at 30° to the horizontal. Obviously, control of attitude, orientation, and rotation are all required, yet this same operation has already been successfully carried out for an articulated loading column of somewhat similar concept.

Mechanical outfitting must continue within the shafts during and after their construction. Temporary doors may be left to facilitate personnel entry and provision of services. The "doors" subsequently have to be concreted closed and must then be able to carry the stresses imposed by hydrostatic and gravity loads and by the global bending of the shaft under wave action. They must, of course, be watertight. Detailing of the closures needs to be carefully worked out by the constructor's engineers in coordination with the design engineer. Particular attention has to be devoted to the upper horizontal joint where bleed water and settlement may allow water to permeate under high hydrostatic head. A secondary injection of grout or epoxy may be utilized. Consideration must be given to residual stresses and differential prestress.

There is typically a large, heavy ring beam or girder to be constructed on top of the shafts. This may be of reinforced and prestressed concrete, as with the Condeep platforms, or of fabricated structural steel, as used at the Ninian central platform. In either case, tolerances, level, and distances must be controlled to within a very few millimeters. Since at the time of their installation the structure will probably have ballast distribution different from that at the time of deck mating, causing bending in the structure as a whole, correction factors must be calculated and applied to the different inclinations and hence distance apart of the shafts.

The base structure and the net forward speed will be significantly reduced or even prevented. In the case of the Statfjord B platform, the tow had to be aborted and rearranged. Use of two pusher tugs then proved far more effective.

Control of the structure within narrow channels has been effectively carried out by having one or more lead boats, auxiliary boats on each side, and either pusher tugs at the stern or small tugs headed the opposite direction and being pulled astern. For such a tow, a temporary control platform is mounted. A generator must be provided for power and lights, with fuel for the trip, freshwater supplies, radios, and navigation gear. A life raft is provided, plus fire extinguishers.

The Statfjord platforms were towed about 100 km to their deep-water mating site; the Ninian central platform was towed only 30 km, but the Andoc substructure was towed from Rotterdam to Bergen, Norway.

A principal environmental concern en route is that of currents, which will act on the very large submerged mass. Winds are of concern only because of their effect on the boats and their ability to hold or change heading.

The deck mating site should be selected to provide adequate depth to enable the substructure to be almost fully submerged by ballasting yet provide good protection from winds and good mooring positions. An ideal location would be at the head of a relatively narrow and deep inlet which is protected from high winds.

Moorings are installed both in the seabed as necessary and onshore, the latter facilitating the needed adjustments in lengths of line as the structure is ballasted down and deballasted up. A second set of moorings is established in order to permit mooring of the deck-barge complex upon its arrival.

While the substructure is being thus constructed, the integrated deck is fabricated. This may be done in a shipyard, on girders or trusses spanning a graving dock or overhanging a trestle (jetty). These support conditions will be different from those that will be acting when the deck is finally mated on the substructure, and hence deflections and twist deformations must be calculated and accounted for. These can amount to as much as 100 to 150 mm in extreme cases and affect the equipment and piping as well as the girders themselves. Most recently, therefore, pillars have been constructed in shallow water near the shipyard. The pillars are short segments identical in configuration and relative location to the top of the shafts of the substructure, enabling erection of the deck under exactly the same support conditions as it will have when mounted on the substructure. It is therefore essential that these pillars be correctly located, with a tolerance in plan of only 20 or 30 mm and in level of 10 mm.

The deck structure will typically start off with large, heavy structural steel ring girders, perhaps designed to be filled with concrete later, after the deck has been finally mounted on the substructure. Then very heavy steel trussing or plate girders are built up with prefabricated sections. Because of the spans and loads involved, plate thicknesses often run up to 100 mm. Welding procedures will probably require both preheat and postweld heat treatment. Since these joints are subjected to cyclic loading, fatigue considerations will probably require grinding of the weld profiles and checking not only of the welds by NDT but also of the adjacent heat-affected zones for hardness.

As the support structure is built up, various modules of equipment are installed. Scheduling is obviously a very complex matter, since equipment modules take time to assemble, yet the support structure must also be completed. Careful planning will enable equipment modules to be lifted over the support structures, lowered, and in some cases skidded to final location.

The deck, now supported on several pillars or other temporary supports, is a very large structure which will undergo expansion and contraction with the temperature and will experience rotations as weights are installed. Therefore, it is usually supported on heavy laminated neoprene and steel pads, allowing shear deformation and rotation.

At the same time, the deck must be adequately secured against a windstorm. Stops must be provided so that the structure can in no way slide off. Wind forces can reach 300 to 500 tons or more under a severe storm. Provisions must also be made for jacking if the combination of temperature cycles and wind cause the structure to "crawl" sideways. Bearing pads should be regularly inspected visually.

During this period, a great many people will be working in confined spaces and with possibly conflicting demands for scaffolding, cranage, ventilation, lighting and access. Since the deck is almost always on the critical path and the labour costs are high, careful planning and scheduling can produce significant savings.

Appropriate provision must be made for fire protection. Also since the work is over water, either floats or nets should be installed to protect anyone who falls. Proper provision must be made for grounding of electric welding equipment. Welding cable insulation should be maintained in good condition and repaired or replaced when damaged by abrasion.

Housekeeping and cleanliness, along with proper lighting, will have positive benefits on safety and efficiency. Cleaning of insulation, sandblast sand, and debris is especially difficult but, of course, is no different in nature than with large ship construction. Some operations will require protection from rain and wind; fireproof canvas, plastic, or similar shelters may be required.

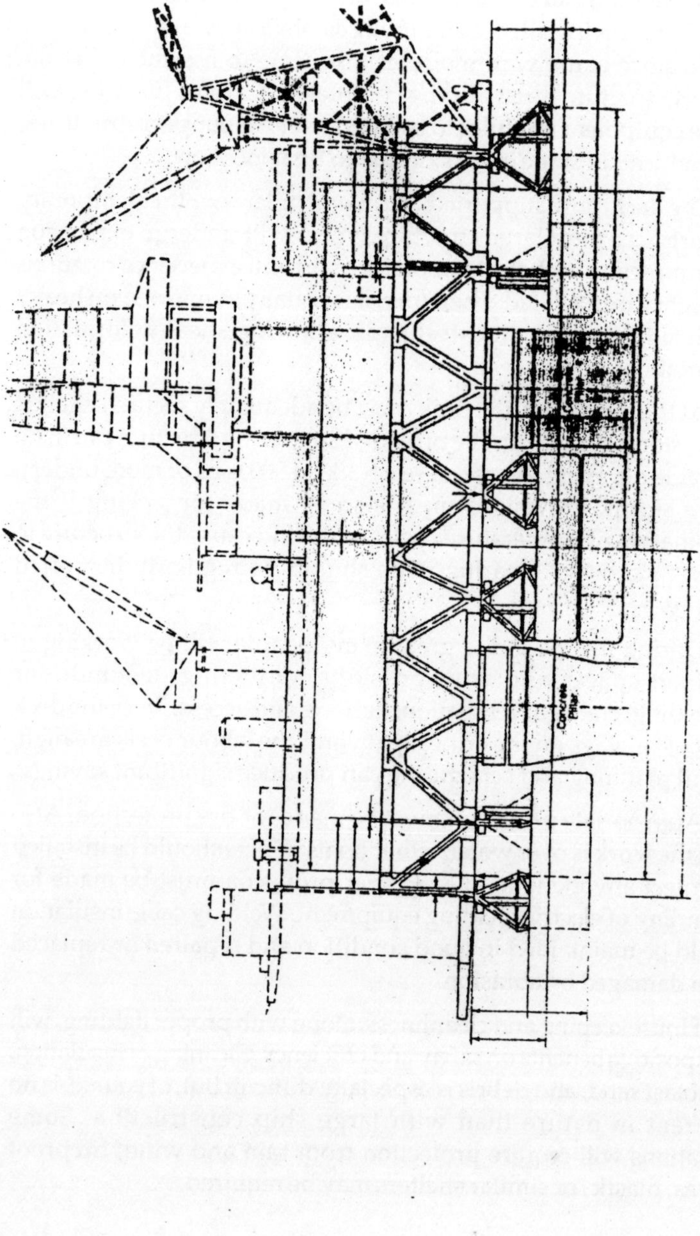

Figure 2.9: Arrangement of Barges for Transport of Statfjord B Deck

Detailed weight control of the superstructure must be maintained, as it must for the substructure, with the added emphasis that weights on the deck are high above the center of buoyancy and hence will have a disproportionate effect on the stability. For example, to counter an extra 100 tons on deck may require adding 1000 tons of solid ballast in the base raft.

When the deck is ready for transfer, it is essential that the weight and its distribution be accurately known, since the deck structure will be transferred by floating equipment. This structure may differ from the final configuration, since one or more modules may have been delayed or postponed. Temporary materials and equipment may be on board and must be accounted for.

The deck is now to be lifted off the pillar supports by means of large barges. These "barges" may be conventional offshore barges, halves of surplus tanker hulls, or special pontoon combinations designed to float in under the deck, between the pillars. They are then deballasted to rise up under the deck, so that sea-fastening

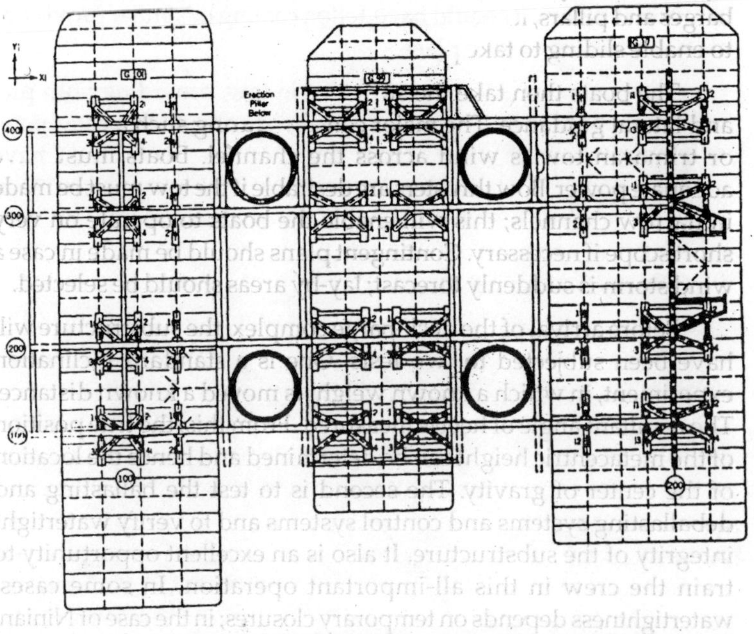

Figure 2.10: Arrangement of Barges for Transport of Statfjord B Deck

supports may be affixed between the barge and the deck girders at appropriate lift points. This operation requires dead calm water.

A careful analysis and evaluation has to be made regarding where to use fixed connections, where to use pinned connections, where to allow flexibility, and where to use ties or struts. With fixed connections, high moments will be developed in the deck trusses or girders, requiring evaluation. The two or three barges, together with the deck, will form a complete structural system and must be analyzed as such, under both static conditions and the dynamic conditions (waves) expected during this particular movement.

When all barges have been properly ballasted, the sea fastenings connected, and the weather forecast is favorable, the barges are progressively deballasted until just before liftoff. Load cells may be used to verify forces in key members. Then the three barges are further deballasted, raising the deck clear of the pillars.

During this operation, the barges are moored to the pillars by taut lines of sufficient length to accommodate the vertical movement. Nylon lines may be used to absorb shock. If blocking is used between barges and pillars, it should have Teflon coating or similar provisions to enable sliding to take place.

The boats then take the structure in tow, providing both pull and lateral guidance. The major concern during such a catamaran or trimaran tow is wind across the channel. Boats must have adequate power. Bow thrusters are desirable if the tow must be made in narrow channels; this will enable the boats to operate on very short scope if necessary. Contingent plans should be made in case a wind storm is suddenly forecast; lay-by areas should be selected.

Before arrival of the deck-barge complex, the substructure will have been subjected to two tests. One is a standard inclination experiment, in which a known weight is moved a known distance. The resultant angle of heel is measured. From this, the true position of the metacentric height can be determined and hence the location of the center of gravity. The second is to test the ballasting and deballasting systems and control systems and to verify watertight integrity of the substructure. It also is an excellent opportunity to train the crew in this all-important operation. In some cases, watertightness depends on temporary closures; in the case of Ninian, several hundred plugs were used to close the holes in the perforated breakwater. These must all be double-checked to ensure they are secured tightly; the failure of only one could cause the loss of the entire platform.

The substructure is then submerged in a series of steps, checking weight of added ballast water vs. resultant draft.

In a number of cases, compressed air has been introduced into the main cells during deep submergence in order to provide additional safety against implosion due to the high hydrostatic forces. It is recommended by both the Federation Internationale de la Precontrainte (FIP) Recommendations for Concrete Sea Structures and the American Concrete Institute ACI 357 State of Art Report on Concrete Sea Structures that the structure have a safety factor of at least 1.05 in the event of loss of air; thus the use of compressed air is determined by the need to further raise the safety factor.

Care must be taken to ensure that when compressed air is introduced into a cell, it is done in a step-by-step pattern matching the external hydrostatic pressure during both deep submergence and subsequent deballasting. The air pressure is constant over its full height, whereas the hydrostatic pressure not only decreases due to deballasting but also varies in a triangular diagram along the height of the cell. Uncontrolled air pressurization could result in an outward explosive force on parts of the structure designed only for external load. Compression of air raises its temperature which will later cool back to that of the external water, in turn reducing the pressure. The typical structure is very large. It is not unusual for the compressed air volume to add several hundred tons of air to the total weight of the structure during the filling process.

During the submergence, the structure is subjected to the largest external hydrostatic forces which it will ever experience; hence stresses are at their maximum. Any structural defects are brought into sharp focus.

In the case of Statfjord A, it was an apparent small crack and leak in the diaphragm wall of a cell, which later turned out to be a serious but repairable delamination of 10 × 20 m in size and up to 6 mm in width.

The tragic catastrophe of the Sleipner failure occurred during the test submergence. When at its deepest draft, the diaphragm wall at a star cell (interstice between the larger cells) failed in shear compression. This was followed a few seconds later by a second similar failure and the entire structure rapidly sank, violently imploding at depth. Fortunately the small crew on board was safely evacuated.

This failure was, of course, the subject of intensive investigation which found the specific technical error to be the reinforcing steel adjacent to the throat of the intersection of the two walls. The star cell was subject to the hydrostatic head of the external sea while the adjacent main cells were deballasted, thus developing a high shear which combined with the global compression of deep submergence.

The relevance to construction is that the originally designed long-headed stirrup bars that crossed the throat had been replaced during construction at the request of field personnel by short bars, in order to facilitate installation in this congested zone. While the overall causes include design and institutional failures as well, it is probable that the original long bars would have prevented failure.

The scene is now set for the deck transfer operation. The weather forecast must, of course, be favorable, with minimum wind. All boat traffic in the vicinity is stopped. Water density measurements at depth are made to be sure that calculations are based on actual densities. The substructure is ballasted down so that only 3 to 5 m of shaft extends above the waterline. The deck-barge complex is slowly winched in around the shafts. Clearances are typically only 300 to 500 mm.

Hydraulic winches can control the structure within 20 mm or so, but the lines have elasticity, making them sensitive to even minor surges. Blocking is therefore employed, arranged with hydraulic jacks, to effect the final horizontal control. The blocks are fitted with Teflon so they can slide vertically.

When horizontal positioning has been verified, it is time to deballast the substructure. There will usually be about 1 m of vertical gap. The substructure is brought up just short of contact; all points are checked. Then deballasting continues. When 10 per cent of the load has been transferred, a final position check is made. All remaining tension sea fastenings are now disconnected.

During this stage, a very complex interaction takes place. As the substructure starts to raise the deck, the barges also rise due to relief of weight. They tend to follow the deck up as it is picked up by the substructure. Since the substructure will now have the weight of the deck on it, the height of ballast water in the cells will have to be further lowered, which means that the net external hydrostatic heads are now the maximum which the structure will ever see. This condition typically is the most critical in the structure's life, as far as implosion in concerned.

Finally, the substructure is picked clear of the barges and on up to a safe height. This "safe height" is one selected to balance safety against implosion (with all compressed air removed), stability, and access for further work. The barges can now be pulled clear.

Since the conditions of support of the deck during transport are inherently different than those on the pillars or on the substructure, there will be differential vertical and horizontal deflections and twist. These must be accommodated either in the flexibility of the deck and shafts or else by the use of bearing devices which will permit lateral movement on the bearings. Sand jacks, neoprene pads, and sliding plates are some of the systems used. Whatever is used, positive stops should be provided.

Vertical tolerance adjustment between the deck and the ring beams on top of the shafts has been provided by a number of means. C.G. Doris has employed flat jacks, which permit measurement and equalization of load. After such adjustment, the water in the jacks is replaced by cement grout. Norwegian contractors have employed soft iron pipe sections, thick-walled, which deform under load concentrations equalizing the load onto adjoining pipe sections.

The deck now rests on the shafts by gravity alone. A check must be made to ensure safety in the event of accidental flooding, which would produce a heel. A typical requirement is that the deck not slide off even if a significant heel (*e.g.*, 7½° or even 15°) is experienced due to accident. Such an accidental heel did occur at a later stage with the Statfjord A platform. During testing of the ballast control system, one valve was left open, causing an un-planned shift of water to one side, resulting in an increasingly severe list. Alarms sounded and the workers started to abandon the structure. Fortunately, the supervisor on the platform had the wisdom and courage to descent deep into the utility shaft, correct the valve errors and ballast the structure back to vertical.

The deck is now additionally secured to the shafts by prestressing. Prestressing is especially effective in eliminating cyclic fatigue and in providing safety against uplift under accidental conditions. These prestressing tendons are usually short bars. A system must be used that permits adjustment for initial seating losses, since in a 4-m-long tendon, a seating loss of 6 mm amounts to a loss of prestress of 350 MPa, almost half that input by the jacks!

These prestressing tendons must also be protected against corrosion, since this is one of the most vulnerable zones. In particular, sealing and drainage must be assured to prevent saltwater spray being trapped in anchorage pockets.

During the entire deck-mating operation, careful checks are made of the ballast water in all the cells to detect any unexplained variances in water level or quantity pumped. Even relatively small discrepancies should be checked by visual examination within the cells. The relatively small in-leakage in one of the cells of the Statfjord A platform, referred to earlier, showed only about 300 l/min inflow, but turned out on more detailed investigation to be due to a very large laminar crack in a cell wall, caused by secondary bending of the large caisson as it was subjected to the high hydrostatic forces of the deep submergence. Extensive repairs were necessary.

The Ninian deck transfer was carried out in a different but very ingenious manner. The deck structure, weighing 7000 tons, was constructed on land, at Inverness, Scotland, then skidded onto a single large barge for tow around the north end to Scotland to the mating site near the Isle of Skye. In a sheltered inlet, the deck was transferred to two barges which supported it only along its edges.

Meanwhile the concrete substructure was ballasted down. In suitable weather; the twin barge assembly was brought to the substructure. By means of lifting rods supported from towers on the centerlines of each barge, the deck was raised so that it would clear the top of the single central shaft. The twin barges now were guided past the shaft, straddling it. The lifting rods now lowered the deck onto the shaft.

Temporary supports were flat jacks on neoprene pads. With the change in support conditions, the deflections of the deck trusses also changed, resulting in lateral movements which were accommodated by the pads. The flat jacks were then used to equalize the load at each of eight support points. Then they were pumped full of grout.

During this stage, which may require 2 or 3 months, the platform is deballasted to raise it to its optimum position, where pressures on the cell walls are reduced and where access can be provided to the shafts.

The next operations are those of hookup of all equipment and its testing. These can require many hundreds of thousands of man-hours. While the site of work is usually protected, nevertheless all access must be by water, and the deck is now 30 to 50 m or so above the water. Hundreds of personnel must move on and off at each change of shift. It is necessary to provide boat landings, adequate crew boats, elevators and stairways, lighting, and water and power supply. As soon as possible, the permanent cranes on deck are made operable so they can handle supplies, tools, and fittings on and off as required. Safety of the workers must be assured; this requires life rafts, a safety boat, and suitable temporary railings. Fire protection must be provided.

The structure is now placed under tow to the site. Its displacement may be several hundred thousand tons; Statfjord B displaced almost 7,00,000 tons. Six or more of the largest tugs in the world are typically employed, each being rated near 20,000 IHP and having a bollard pull in excess of 150 tons. These are arranged with towing pendants and retrieving lines as described earlier. For the Troll platform, displacing almost 1,000,000 tons, eight tugs were employed.

Within restricted waters, two stern tugs are also employed, to prevent yaw and to keep the GBS platform fairing true behind the boats. Then, when the tow emerges into the open ocean, the scope of towlines is lengthened and the stern tugs are cast off.

A pilot boat may lead the way during exit from the channel, available to warn other shipping and in some cases to verify clearance across relatively shallow waters. A safety boat may run alongside, for use by supervisory personnel, interboat transfers if needed in an emergency, and to pick up a person overboard.

The GBS platform is outfitted for the tow with a navigational bridge, radios for communications to boats and to the shore base, electronic position-finding equipment, safety and firefighting equipment, quarters for the riding crew, diesel and water supply, and generators for power and lighting. The heliport on the deck may be activated for emergency use. Duplicate radar, gyros, fathometers, and forward-searching sonars are installed, as well as emergency generators.

Dynamic response of the platform's motions must be considered under the action of the design storm. This is usually taken as the 10- or 25-year return storm for the period of year involved. With its relatively low metacentric height, the structure will typically roll and pitch in a relatively long natural period, perhaps 60 s, developing maximal accelerations as high as 0.20 to 0.25 g. The accelerations, of course, affect the deck equipment and require that all attachments be capable of resisting the lateral forces which develop.

It is important to calculate righting moments for the various angles of heel and the heeling forces due to wind and waves, since a typical platform does not have the response of a conventional ship hull and metacentric height is a measure of stability only at very small angles of heel. Usually, a limit of 5° will be placed on the maximum heel during the 10-year storm, assuming the towline is slack or has been dropped.

For large structures and major offshore platforms, the use of temporary buoyancy would appear to be more conservative and appropriate. Temporary buoyancy tanks will have to be relatively large in diameter (*e.g.*, 5 to 10 m) and high (*e.g.*, 40 m) to raise the center of buoyancy without significantly affecting the center of gravity. Therefore, they will usually be made of steel, with heavy ribs, although one study has shown that a hybrid steel-light-weight concrete shell was even more efficient. Temporary buoyancy tanks also give a minor improvement in the moment of inertia of the water plane but unfortunately also increase the displaced volume, so that the net effect on stability is not usually significant.

Temporary buoyancy tanks give their major contribution by raising the center of buoyancy. Therefore, they do not have to be symmetrically placed around the structure. For example, one such tank can be installed on one side only, if that is desired, and verticality maintained by selective ballasting.

These tanks must be attached to the structure at some point on the enlarged base caisson. These attachments are subject to the dynamic cyclic forces of the waves during the tow, which can lead to low cycle fatigue, especially in the corrosive environment.

Prestressing of the tanks to the base appears one good means of attachment, using post-tensioned unbonded tendons prestressed from the top of the tank at a stage when that top is above water. The tendons can then be released after installation, again working from the top of the tank. Alternatively, they can be cut by explosives, hydraulic shears, or underwater burning. To release such tanks, they should be ballasted to neutral buoyancy and then released so that they do not rise up suddenly. A towline to an attachment at the center of rotation of the tank will prevent the tank from rotating during release and rise, with possible damage to the shafts, deck, or tank.

Temporary buoyancy tanks were employed on 10 caissons of the offshore terminal at Hay Point, Queensland, Australia. During the installation of the last caisson, one connection failed due to corrosion-accelerated fatigue; fortunately, this happened when the structure had almost touched down, so there were no serious consequences.

Returning to the structure at the side, it is being held in position by the tugs (or mooring lines) and being ballasted down by pumping or flooding in of seawater. All events and conditions are being monitored in the control room. As it nears the seafloor, the rate of descent is slowed to allow the water to escape.

The concrete caissons for the offshore terminal at Hay Point, Queensland, Australia were all installed within 0.5 m by the use of anchor winches on board, with lines leading either to preplaced seafloor anchors or to previously placed caissons. Similar procedures were followed with the four concrete caissons for the Tarsiut Caisson Retained Island in the Canadian Beaufort Sea.

At touchdown, the dowels plow into the soil to sufficient depth to prevent the caisson from further lateral movement. Bending stresses, converted from strain gauges on the dowels, are read out in the control room. The short-range, high-frequency echo sounders on each corner give the distance between base slab and seafloor.

The skirts now engage. These have been designed not to buckle even if they hit a subsurface boulder; they will displace it laterally through the soil. Ballast continues to be added until the desired penetration is achieved. The initial penetration rate is kept slow, *e.g.*, 150 mm/h, to avoid overpressure in the skirt compartments and consequent piping. Once skirts are well embedded, the rate of penetration may be increased to about 1 m/h. Meanwhile, water is being vented out from the skirt compartments.

Eventually, full penetration is achieved and the structure comes to rest on the base slab or on concrete skirts or sills purposely designed to halt further penetration and thus prevent excess local pressure on the slab. Without such arresting, the structure will continue to penetrate until it bears directly on the slab. High spots of clay will be displaced and squeezed out. Even a surface boulder will be forced down into the soil. A high spot of sand, however, may give very high local bearing resistance, thus overloading the slab. Pressure cells on the base slab and strain gauges may be used to monitor slab-foundation interaction.

If the skirts do not penetrate the required distance, even when the structure is fully ballasted, then the water trapped inside the skirt compartments may be bled off into the utility shaft at an underpressure, thus increasing the effective force causing penetration. Such selective ballasting and underpressuring allows very accurate control of level. The degree of underpressure must not be so great as to initiate piping under the skirts. Use of underpressure has proved to be essential for achieving penetration of long skirts such as those on Draugen, Gullfaks C, and Troll. For the Gullfaks C platform, 22-m-long concrete skirts penetrated the soft clays with interbedded lenses of sand. By reducing the pressure inside the cells, the necessary driving force was developed. A large-scale field test demonstrated the enhanced effectiveness of cycling the internal pressure.

For other platforms in similar soils, where penetration of the entire base raft is required, it has been proposed to progressively remove the soils from within the skirt compartments by airlifts, eductors, or pumps.

Once the structure has been penetrated to its designed embedment, the spaces remaining beneath the base slab are often filled in order to ensure equal bearing on the soil. Weak grout fill has been used in many cases. Thixotropic admixtures can be added to the grout to reduce segregation and prevent excessive intrusion of the grout into interstices of rock and grout escape through minor openings into the sea. For example, for offshore caissons in Australia which were seated on a prepared base of crushed rock, tests showed that normal grouts were much too fluid and penetrated too far into the rock, whereas a thixotropic grout gelled as soon as the resistance to flow increased.

Grouts for underbase fill must always be placed under a very low head to avoid development of piping under the skirts and to avoid lifting the caisson. Use of a gravity feed through an open hopper at the correct elevation needed to just overcome the hydrostatic head by about 15 psi (0.1 MPa) is the best procedure. Provision must be made for venting of the displaced water. If pumping is employed, this must be at low pressure, with positive vents to prevent overpressurization.

When the void under the base is thick, for example, 0.5 to 2 m, as it is in many North Sea installations, then means must be taken to prevent excessive heat of hydration from developing in the underbase grout. Some full-thickness tests showed that the temperature with normal cement mixes could rise to 100°C and more, creating undesirable strains in the base slab.

Finally, each installation needs to be evaluated as to what is desired. In many cases a low modulus of elasticity is desirable. Strengths need be little better than that of the foundation soils: 1 to 2 MPa may be a reasonable value in many cases. Bleed should be minimal. In the North Sea, underbase grout mixes have been developed which are a slurry of cement and seawater, with a foaming admixture such as sodium silicate, from 4 to 10 per cent by weight of seawater, along with retardation and stabilizing admixtures. This presents minimum logistics problems and develops a low-strength, low-modulus mix that flows well and tends to fill up underbase cavities.

For caissons in relatively shallow water or where strong bottom currents are present, underbase grout must be protected from erosion either by shirts or an antiscour stone mattress. A higher-strength grout and quicker set are required. Cement-based grout fills were flowed into the underbase gap of the 66 gate support caissons of the Oosterschelde Storm Surge Barrier and under the caisson piers of the Oresund Bridge. Antiwashout admixtures should be employed.

In cases where the platform will later be refloated, for example, an exploratory drilling structure for the Arctic, or where a permeable underbase fill is desire, properly graded sand may be slurried into the underbase spaces. The sand slurry is flowed in through piping, either internal or external, having minimum bends; those bends that are necessary should be of large radius. The sand tends to drop out of suspension and build up a little dam; this causes a slight rise in pressure, and a rivulet of sand slurry breaks through to create new fill and successively new dams. Spacing of discharge points is usually 3 to 5 m on center, assuming a 30-cm depth underbase void is to be filled.

Sand flow methods have been thoroughly developed by both Danish and Dutch engineers for use in their submerged tube and harbour construction. Offshore, they have been successfully applied under Arctic exploratory drilling caissons. In all cases of underbase filling, whether by grout or sand, the feed should be at low head and well controlled to prevent raising the structure by excess pressure.

The next item of work is that of scour protection, if required. Skirts which penetrate are one means of scour protection; however, these are not always practicable due to soil conditions or draft constraints.

Rock may be dumped around the periphery of a caisson, using a long discharge tremie from a rock hopper barge. This system was developed by Dutch dredging contractors for covering submarine pipelines in the North Sea. Uncontrolled dumping from barges caused significant impact damage to the concrete piers of the Oosterschelde Storm Surge Barrier.

In sandy soils a filter fabric of some sort is desirable under the stone to prevent the sand from leaching through the interstices in the stone. Fabric mats can be preattached to the edge of the base slab and rolled up along the sides of the lower walls. After landing, they may be cut loose and spread out radically from the base. To hold their outer edges, divers may place sandbags or drive a headed steel pin through them to anchor them in the sand. This procedure was employed on the Ekofisk caisson. The filter fabric was the covered by dumped rock. Shortly after completion of this work, the structure encountered a major storm. The scour protection proved to be fully adequate.

Filter fabric for this application should be heavily reinforced with nylon or even stainless steel wire. Such a heavily reinforced fabric, sewn together with a finer mesh geotextile, has been used extensively in coastal applications in The Netherlands.

Dumped rock may be impracticable where there are flow lines and other mechanical fittings near the base of where it is impracticable to locate such a vessel in the vicinity of the platform, keeping in mind the many other activities which must be carried out during these first few critical months. The primary effort must, of course, be the spudding in and drilling of the wells.

Hence, one solution is to place prefabricated mats made up of heavily reinforced filter fabric to which concrete blocks have been preattached. Such mats can be placed by derrick barge, using spreader beams; large mats would have been laid out and stacked on a barge for easy handling. These were used to provide scour protection for the caissons of an offshore terminal in Queensland, Australia. They performed successfully through a major cyclone shortly after installation. Such mats can also be affixed to the periphery of the GBS base and stopped off up along the sides. After seating of the GBS, they can be cut loose and laid out on the seafloor with the aid of cranes or workboats. Such mats could probably not stand the action of the sea on a long open sea voyage and hence would not be attached until at or near the final installation site. They appear well suited to use in the Arctic, where they might be attâched during the wait for ice to clear at a specific installation site.

Articulated concrete block mats such as those used to protect offshore islands in the Arctic, have been planned for use on a large concrete caisson structure for the Arctic. They are to be transported on the deck of the structure. On arrival, they would be attached to the base and laid out over the peripheral soils.

The final construction operation at the site is to install the conductors. These will usually have been transported with the GBS structure and already set in the conductor guides. In the typical platform base, conductor sleeves will have been concreted in the base slab and then sealed against waster entry during transportation.

While a double set of reinforced neoprene pile-closure sleeves might be utilised to seal these conductors, most closures to date have been a 1.5 to 2-m plug of unreinforced concrete, which is later drilled out by the drill rig.

The conductor sleeves, penetrating the bottom slab, have both internal and external mechanical shear strips or keys to prevent any possibility of shear failure, as this could cause the loss of the platform during tow. When the sleeves are later drilled out, the drilling water level inside the drill shaft should first be equalised with the external sea level. Once the drill penetrates, any significant head differential will either cause the foundation soils to flow in or piping to take place to the outside.

In the case of Beryl A, when the head inside had been pumped down about 15 m to permit installation of fittings in adjoining cells, as soon as the drill penetrated, several hundred cubic meters of sand flowed into the shaft. Piping in from under the outer cells occurred. Satisfactory support and filling was restored only by extensive underbase grouting and external filter plus rock protection.

For subsequent structures, skirts have been provided around the drilling shafts and water heads have been kept balanced during this operation. Skirts around the drilling shafts will also limit the flow of drilling mud which may spill over from the wells into any adjoining underbase voids.

The conductors are now driven and drilled down to the prescribed penetration, following which the surface pipe is run and the wells drilled. The annuli between well casing and surface pipe and often that between surface pipe and conductor are cemented.

The Movement of Spuds and the Relocation of Platforms

In some installations at sites with very weak soils, it may be necessary to augment shear transfer from the base to the foundation in order to prevent sliding. This need can arise with platforms in the Arctic where very high lateral ice forces must be resisted yet skirts are not practicable due to draft restrictions for transport, or where there is inadequate weight of the caisson structure, even when fully ballasted, to drive the skirts into the soil.

Spud piles were proposed for the Sohio Arctic Mobile System (SAMS). These would be driven after founding of the platform. They would be jacked or driven through steel sleeves to engage firmer soils such as partially ice-bonded dense sands at some depth below the seafloor. A typical spud might be 2 to 2.5 m in diameter, with wall thicknesses up to 100 mm, designed to penetrate 12 m. It will extend a further 20 m up into the spud well; 10 to 20 such spuds might be used on a platform. The spuds could, of course, be jacked in, in the same manner as the legs of a jack-up rig. The internal plug can be broken up by jets, which are permanently attached to the

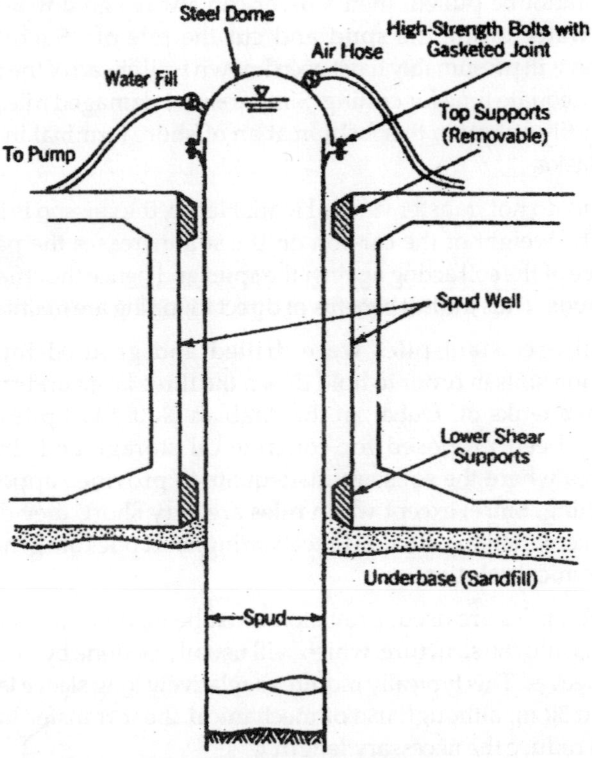

Figure 2.11: Spud Removal

sides of the spud. Alternatively, the spuds can be driven in with a large vibratory hammer or even a diesel or steam hammer. Penetrations are very short as compared to piles, and hence installation time should be minimal.

If subsequent removal is required due to the desire to relocate the platform, the spuds can be pulled up using the following procedures.

Spud piles can also be used effectively in seismic areas where shear transfer to the foundation soils is required yet the mass of the structure is limited and hence there may be insufficient friction between the base and soil. At the same time, the low mass reduces the seismic forces. Spuds can then be used to supply the needed shear resistance.

If a spud has become buckled or distorted so badly by overload that it cannot be pulled, then a diver can always go down in the heated water inside the spud and cut the pile off. Such major distortion will presumably have been known well ahead of the move, giving adequate time for cutting. Similar-sized damaged piles have been readily cut off in this fashion at an offshore terminal in Cook Inlet, Alaska.

Spud do not transfer vertical load. Hence, the caisson is free to settle. The weight of the caisson on the soil increases the passive resistance of the soil acting against the spud and hence the efficiency of the spuds. Thus the full benefits of direct founding are maintained.

Uplift-resistant piles were drilled and grouted into the foundation soils in order to hold down the three large underwater oil storage tanks off Dubai, in the Arabian Gulf. Skirt piles have similarly been proposed for concrete oil storage and drilling structures where the soils are inadequate to provide support for overturning. Since, except when piles are very short, they do not work in conjunction with the direct bearing, skirt piles are generally an economical solution.

When piles are used, provision must be made to transfer the pile loads into the structure, which will usually be done by grouting within sleeves. This typically requires a relatively long sleeve length, say, 20 to 30 m, although use of mechanical shear transfer lugs or ribs will reduce the necessary length.

Skirts may also be used to develop supporting capacity for the structure. For example, 35-m-long skirts were used on the Troll platform, which is located in over 300 m of water depth, with extremely soft surficial soils. The skirts were designed to develop both skin friction (cohesion) and end bearing in the soils, thus minimizing long-term settlements due to compressibility. Such a solution could also be applicable to the similar conditions in the Navarin Basin of the Bering Sea.

The penetration is often able to be attained by ballast weight alone, provided provision is made to vent the trapped water and semifluid mud. This may be augmented by reducing the water pressure within the skirts, so that the ambient hydrostatic head, acting on the horizontal surfaces, forces the structure down.

Alternatively, techniques previously used on the caissons for deep bridge piers may be adopted, namely, jetting and removal of

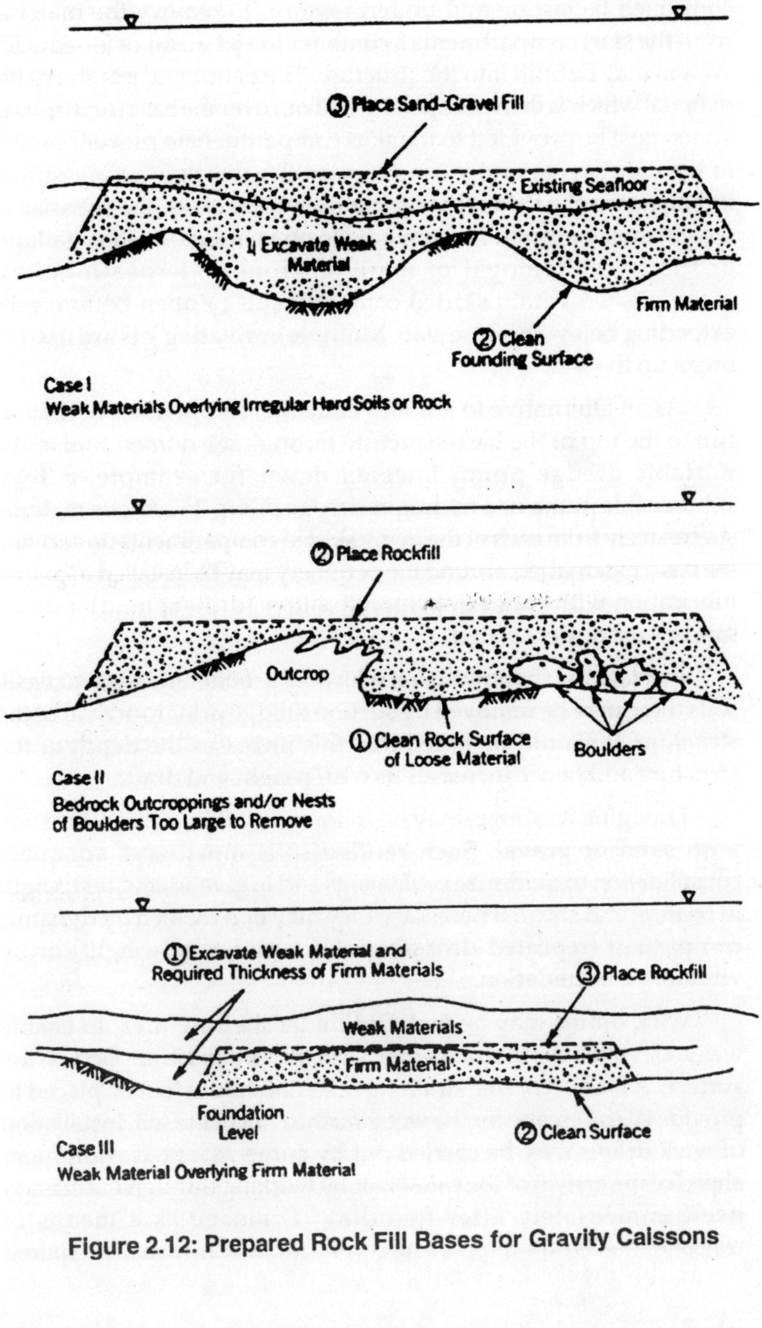

Figure 2.12: Prepared Rock Fill Bases for Gravity Caissons

the soil from under the base while increasing the driving weight by controlled ballasting and underpressure. To remove the material from the skirt compartments a combination jet-airlift or jet-eductor system may be built into the structure. The peripheral jets slurry the material which is then pumped up and out over the base for disposal. Vents must be provided to the skirt compartment to prevent piping in the case severe under- or overpressure develops as a result of imbalance in the flows. In weak soils such as clays, the increase in weight must be carefully monitored to prevent sudden shear failures of the soils. Removal of material from underneath is best accomplished within skirted compartments or open-bottom cells extending below the base slab. Multiple or rotating jets are used to break up the soil.

As an alternative to jets and eductors, an open sleeve may be run to the top of the base structure through the domed roof and a portable dredge pump lowered down–for example, a Toyo submersible pump or a hydraulic dredge pump–to remove material successively from each of the many skirted compartments underlying the base. External jets around the periphery may be installed to permit lubrication with water or bentonite slurry (drilling mud) if dense sands must be penetrated.

In relatively shallow water, with only a moderate depth of weak soils these may be removed by suction dredge prior to arrival of the structure.. It should be noted that this increases the depth of the structure and hence increases its cost, weight and draft.

Dredging as above, may be followed by filling the excavation with sand or gravel. Such refilled soils must have adequate consolidation to minimize settlements and to give adequate strength in bearing and shear. If necessary they may be densified by dynamic compaction (repeated dropping of a very heavy weight) or by vibratory consolidation.

Wick drains may be installed under the base area, to enable weak, claylike soils to drain as the structure weight is applied on the surface. A blanket of free-draining material should first be placed to provide lateral escape for the water pushed out of the soil. Installation of wick drains may be carried out by barge-mounted equipment ahead of the arrival of the caisson or by working through the caisson itself immediately after founding. Drainage as a means of consolidation is time dependent; several months are usually required for full strengthening to resist design loads. This may be acceptable in some regions of the Arctic and sub-Arctic where the design ice load is not expected until 3 to 5 months after installation.

Embankments may be placed at the site in the season preceding installation, while the GBS is being built. Especially if these are extended over an area larger than the structure's footprint, then they will help to consolidate the clays and increase their shear resistance, as well as provide external berm resistance to shear failure in the soil. Of course, external berms of stone can also be installed after the caisson has been founded in order to increase the effective resistance of clays to bearing failure and sliding.

Multiple sand piles or stone columns may be driven in a closely spaced pattern under the area on which the structure is to sit. Such sand piles, if closely spaced, are effective in providing direct bearing and shear resistance. If a blanket of permeable material has been placed over the seafloor, then these sand piles will also act as drains when the structure's load is imposed.

Chapter 3

The Principal Materials for Offshore Structures

Introduction

The principal materials for offshore structures are steel and concrete.

Steels for structures or components that will be utilized in cold climates must have adequate toughness at those temperatures. Welding materials are equally critical in assuring proper strength and ductility in service. The weld metal is to be compatible with the base material as regards heat treatment and corrosion. Crack-opening displacement tests or other fracture mechanics tests are normally to be conducted for selection of the welding consumables.

High-strength bolts and nuts, when used as structural elements, should have Charpy V-notch toughness values as required for the structural steel members being connected.

The spatial relationship of structural elements is critical for offshore structures which are to be assembled in the field or where major components are to be mated. API-RP2A provides specific tolerances final fabrication. For jacket and deck section columns, in any plane critical to field assembly, the horizontal distance from the centerline of the adjacent columns should be within 6 mm (1/4 in.)

of the design dimension. The same tolerance should be applied in the other planes to working points on the outside of the columns.

Angles between corner columns should be within 1 min of the design angle. Diagonals of rectangular layouts should be within 18 mm (3/4 in.) of each other. Alignment of jacket columns should be maintained within 6 mm (1/4 in.). For jacket and deck section bracing, all braces should be within 12 mm (1/2 in.) of the design dimension. The deck beams and cap beams at their ends should also be within 12 mm (1/2 in.) of design position.

The assembly of a jacket frame, often having a spread at the base of 50 m or more, places severe demands on field layout and survey and on temporary support and adjustment bracing. A tunnel laser has been utilized to provide accurate levels and alignment on some recent platforms. Such large dimensions mean that thermal changes will be significant. Temperature differences may easily be as great as 20 to 30°C from before dawn to afternoon, and half that between various parts of the structure, resulting in significant distortion (*e.g.*, 30 to 40 mm). On platform Cerveza, members were cut and shaped to the cold-side dimensions in the morning and then held for the midday sun to tightly fit metal to metal for welding.

Steel is subject to a variety of corrosion phenomena: atmospheric corrosion, splash zone corrosion, crevice corrosion, etc. Recently, many steel structures in service in seawater have been corroded by microorganisms spawned by the interaction of aerobic and anaerobic bacteria.

Painting and coating of the steel members, where specified, should be carried out as far as practicable in the shop, under appropriate conditions of humidity and protection from extremes of weather. The joint surfaces should, of course, be masked to permit welding. Field coating of the joints and touchup of shop coats should be done only when the surfaces are dry and at the proper temperature. In some locations, portable tents or other protection will have to be provided. Heaters and/or dehumidifiers may be required.

It is extremely important that surface preparation be thorough and in accord with the specified requirements. The offshore environment will quickly degrade any coatings placed on damp steel, or over mill scale, or rust.

DNV rules require that the provisions for coating include:

1. A description of general application conditions at coating yard.
2. Method and equipment for surface preparation.
3. Ranges of temperature and relative humidity.
4. Application methods.
5. Time between surface preparation and first coat.
6. Minimum and maximum dry film thickness of a single coat.
7. Number of coats and minimum total dry film thickness.
8. Relevant drying characteristics.
9. Procedure for repair of damaged coating.
10. Methods of inspection–for example, adhesion testing and holiday detection.

Surface preparation and application of coating should be carried out when the surface temperature is more than 3°C above the dew point or when the relative humidity of the air is below the limits recommended by the coating manufacturer. Coatings are usually applied to steel in the splash and atmospheric zones and to internal spaces which are exposed to seawater. In the case of sealed internal spaces permanently filled with seawater, corrosion inhibitors may be added to the water prior to sealing.

Sacrificial anodes or impressed current cathodic protection are normally used to protect steel below water. Anodes must be carefully installed in accordance with the specifications to ensure that they cannot become dislodged during transport, launching, installation, pile driving, and service. An adequate electrical connection between sacrificial anodes and the steel structure is essential. Impressed current is more effective because it is less likely to be shielded, but requires continued monitoring and adjustment. It is prohibited in closed spaces or where water flow is restricted because of the possibility of generation of hydrogen.

Coatings may be applied to members which will be underwater in service in order to minimize the requirements for cathodic protection, provided the coating has adequate resistance to cathodic disbandment. Zinc-based and aluminium-based alloys have been applied by thermal spray.

In the splash zone, additional protection may be provided by means of monel wrap, copper nickel, austenitic stainless steel, or carbon steel plate wrap or simply by allowing an added steel thickness in order to provide for some corrosion. Allowances of 0.1 to 0.3 mm/year are made, with the higher values Prestressed and reinforced concrete has been used for over 25 large offshore platforms, mostly in the North Sea. Concrete lends itself to the gravity-base box caisson-type of structure, especially when developing a large field and when offshore storage is required. Recently, a large concrete platform designed to resist iceberg impact as well as North Atlantic storm waves was installed off Newfoundland. Smaller oil production platforms have been installed off Australia and Brazil. Structural concrete has also been used in the Arctic Ocean north of Alaska and Canada, as mobile exploratory drilling structures. Two concrete floating platforms are in the northern North Sea.

Smaller concrete box caissons have been used for over 100 piers for overwater bridges in the waters surrounding Denmark. Concrete box caissons have been extensively used for breakwaters, seawalls, loading terminals, and quays for the berthing of ships. Large concrete barges are in service off Indonesia and West Africa. Concrete is also used in conjunction with structural steel in hybrid and composite designs. Cement grout is used in conjunction with steel platforms, to bond the piles to the skirts and jacket legs.

Structural concrete itself is a composite material consisting of aggregate, cement mortar matrix, reinforcing and prestressing steel. Structural concrete should conform to the best practices of concrete construction and codes as set forth in building codes and recommended practices for bridges and marine structures, as applicable. As in the case of steel offshore structures, the harsh environment and the special loading combinations and operating requirements make it necessary to supplement such general documents with recommended practices and rules for marine and offshore concrete structures. The principal recommended practices used are listed in the Introduction and in the Bibliography. Excerpts relating to construction are presented in the appendices.

Structural reinforced and prestressed concrete has traditionally employed natural sand and gravel aggregates, having a unit weight of about 2300 kg/m^3 (145 lb/ft^3). More recently, especially for Arctic platforms, where shallow water limits the draft, and for floating structures, structural lightweight concrete has been employed, having a unit weight of about 1830 kg/m^3 (115 Ib/ft^3). Both these unit weights are wet weights, and do not include the reinforcing steel.

Structural concrete as a whole and its individual components must be designed to work together in effective composite action. It must be durable under exposure to the sea and the air. Emphasis is placed in design and construction on quality assurance to assure a long life with minimal maintenance.

The Properties of Concrete Mixes

For modern offshore construction, the desired properties are often complex, demanding, and occasionally conflicting to some degree, thus requiring development of an optimal solution. Compressive strength has historically been the controlling parameter by which concrete quality has been measured. We now know that it is not necessarily an accurate or adequate measure of other qualities. Recent advances in concrete technology have led to significant increases in concrete compressive strength, and the trend continues.

Tensile strength determines the onset of cracking and shear strength and influences fatigue endurance. Reinforced and especially prestressed concrete shows excellent fatigue endurance in the air, as long as the concrete is not cycled into the tensile range to a greater level than half its static tensile strength or into its compressive range more than half its compressive strength. These limits are normally met in practical design.

When submerged, conventional concrete shows a reduction of fatigue endurance, apparently due to high pore pressures generated within the microcracks. Interestingly, structural lightweight concrete using modern high-strength, lightweight aggregates and special cementitious admixtures (microsilica shows no such reduction).

Fortunately, even the reduced S-N curve or equivalent for submerged prestressed concrete structures of normal-weight concrete is adequate for water depths of current interest. For the greater depths envisaged in the future, use of microsilica is definitely indicated.

Permeability is an extremely important property. Low permeability to seawater and chlorides is desirable to minimize the occurrence of corrosion. Low permeability to water and especially the inclusion of entrained air is important to the prevention of freeze-thaw damage in cold environments. The use of cement having a moderate tricalcium aluminate (C3A) content is beneficial in that it combines with the seawater chloride ions to form an insoluble compound that blocks the pores. Permeability in concrete occurs primarily along the interfaces between the aggregate and the cement paste matrix. It can be minimized by selection of a mix with minimum bleed by the use of aggregates having surface characteristics that promote physical or chemical bond and by adopting a low water-cement ratio.

Both the quantity and the quality of entrained air (its pore or bubble size and its spacing) are important to ensuring durability in low-temperature areas. Because the concrete is being used in the, marine environment, it has the potential for water absorption and saturation; when such a condition exists, the number of freezing and thawing cycles to cause freeze-thaw disruption is significantly reduced. The sea itself, rising with the tide and the waves splashing over the cold concrete surfaces, thaws the concrete, thus drastically increasing the number of cycles of freezing. Verification of proper air entrainment requires petrographic examination of hardened concrete test specimens.

Abrasion resistance formerly was felt to be determined solely by the hardness of the aggregate. Now it is recognized that the strength of the cement paste and the bond with the aggregate are also prime factors. The use of condensed silica fume is especially advantageous in critical applications.

Sulfate resistance is almost a nonproblem in seawater when rich and impermeable concrete mixes are employed, except in areas having extremely high sulfate contents (the Arabian Gulf, for example). For such cases, the addition of pozzolans is especially useful in both reducing permeability and in combining with the free lime to reduce the chemical attack of the sulfate ion. For structures in freshwater containing significant sulfate content (greater than 1300 ppm), the use of pozzolanic admixtures, such as fly ash, should be used and the C3A should be limited to less than 4% maximum.

Resistance to petroleum compounds and crude oil is provided by normal high-quality concrete. Resistance can be enhanced by the inclusion of pozzolan (*e.g.*, fly ash) in the mix. This also enhances the resistance to the anaerobic sulfate/sulfide-producing bacteria such as *Theobacillus concretivorus* which is present in oil.

A high modulus of elasticity increases stiffness; a low modulus enhances energy absorption and ductility. The modulus of concrete is proportional to the square root of the strength.

Creep has long been considered an undesirable property because it reduces the effective prestress and produces permanent deformations such as sag. However, it also is beneficial in enabling concrete to adjust to sustain locally concentrated loads, thermal strains, and differential settlement. Thus it often reduces the onset of cracking.

Fire resistance is important in the structural portions embodying operating facilities such as utility or riser shafts or where hydrocarbons may be accidentally released. On the typical marine structure, the principal elements of fire resistance which are of interest are spalling, thermal conductivity, and creep at elevated temperatures. Spalling can be limited by the inclusion of reinforcing ties through the thickness.

Bond properties are of special importance when mortar or grout are used to transfer shear from piling to skirt or jacket sleeves. Bond is also of importance in the anchoring of reinforcing steel and of prestressing tendons.

Heat of hydration must often be limited in order to reduce the temperature gradients which later arise when the outer surface cools or when the element as a whole cools but is restrained. Too high a heat of hydration can lead to thermal cracking. Heat of hydration may be reduced by using coarser grind cement, by controlling the cement chemistry (*e.g.*, using blast furnace slag cement), or replacing 20% or more of the cement by fly ash, and by cooling the mix.

Conventional Reinforcing Steel

Conventional reinforcing steel is also called passive steel because it is nominally under a state of zero stress when enclosed in the fresh concrete.

Actually it usually ends up being in mild compression due to shrinkage of the concrete after hardening. The typical marine structure uses very heavy concentrations of reinforcing steel, far more than the usual land structure. To provide adequate space for concrete placement, the bars may be bundled in groups up to four. Special consideration should be given to the design of the concrete mix to ensure that adequate "paste" (cement-sand mortar) is available and that thorough vibration, preferably internal vibration, is employed, to fill the interstices between bars completely. Coarse aggregate size may be reduced.

Cover over the reinforcement is important for durability in the marine environment. Too much cover increases the width of cracks, while too little leads to easier access by chloride and oxygen and in extreme cases to loss of mechanical bond.

Reinforcing steel requires an adequate development or bond length at the end of each bar, in order to be able to develop the full strength of the bar without pullout. Because of the dynamic and cyclic nature of loadings on offshore structures, bond lengths are often greater than those specified for static loads, up to as much as double. The positioning of the ends thus becomes of importance, to ensure that the anchorage is in a zone that is able to develop bond, that is, within a compressive zone. Provision of confinement reinforcement in zones of anchorage and splicing is important. Tie wire for steel reinforcement should be black (uncoated), soft iron wire, never copper or aluminium; otherwise local pitting may ensue. Welded fixing of bars may lead to pitting corrosion and loss of fatigue endurance.

Stirrups are extensively used in offshore structures. It is desirable that the tails of all stirrups be anchored back within the confined core of shells, slabs, and beams, but this is often impracticable. Tying of the free end helps. Because of this problem and the fact that at high loads the concrete under the bend of stirrups crushes, stirrups rarely develop their full yield strength. Because of the bends in stirrups, their fatigue strength is limited. Therefore, headed bars, such as forged or fabricated T-bars, anchored behind the in-plane reinforcement have been developed to provide a more efficient tie and shear-resistant element. These have been extensively employed on recent offshore platforms, primarily for stirrups but also for longitudinal bars as a replacement for hooks, in order to obtain a positive anchorage.

Splices of reinforcing bars may be made by lapping; however, the long splice lengths and the congestion when trying to lap-splice bundled bars raises many difficulties. Mechanical splices are desirable for the larger bar sizes. Welding of reinforcement if used must employ low-hydrogen electrodes. Working drawings must show splice details so that congestion may be evaluated and the weights of steel accurately calculated, the latter for purposes of draft and stability control. Splices of bars destined to work only in compression may, according to the code, be end-bearing. However, as a practical matter, in most cases the reinforcing steel in offshore structures must work in both tension and compression under the variable loading conditions. Therefore, it is considered good practice to design splices to transmit both. If lap splices are used, the bar size should be limited to 32 mm (1¼ in.) diameter and the laps should be well tied at both ends and confined. Splices in adjacent bars must be staggered. Mechanical or welded splices are preferred for the larger-diameter bars.

Corrosion of reinforcing steel is the most common form of degradation in the marine environment. Corrosion protection for the reinforcement has traditionally been provided by selection of the proper concrete mix for low permeability and cement chemistry, along with an adequate cover and good curing.

In recent years, the efficacy of epoxy coating bars for marine structures has been questioned. The experience with straight bars in bridge decks subject to deicing salts has been generally excellent, whereas the experience in marine construction and especially the splash zone has been varied. Epoxy coating has the disadvantage that it prevents adhesive bond, thus giving reduced bond capacity and leading to delamination under severe loading. There have also been a number of cases where the fusion bond of the epoxy coating has been destroyed by saltwater.

Therefore, this author recommends that the use of epoxy-coated bars be evaluated in each specific case of marine construction and not be just generally adopted, as required by some existing codes and guide lines. In any event, the quality of the concrete, and especially its impermeability, is the most important factor in ensuring durability. A practical problem in construction arises when bars of different steel grades or both coated and uncoated bars are to be placed in the same general regions. Clearly identifiable coding must be employed or else all the steel should all be of the highest grade. Clearly identifiable color coding is preferable to the marks stamped on the bars' deformations; the latter are of use in the warehouse but not in the field for installation and inspection.

The supports for reinforcement should be concrete " be" blocks or plastic chairs. Stainless steel chairs may lead to local pitting of the bar in a saltwater environment.

Practicable methods of cathodic protection of embedded reinforcing steel have been developed and applied where serious corrosion is anticipated, *e.g.*, saline water conversion plants and seawater cooling systems for power plants. Cathodic protection by sacrificial anodes can be used underwater: impressed current CP is required for above-water applications.

Bending in the field must follow carefully detailed procedures. Reference is made to the ASME/ACI Code for Nuclear Containment Structures, where provisions for field bending are carefully prescribed. Small tie bars, 10 mm and less, of weldable steel may be field-bent as required.

The Hybrid Steel Structures and Composite Construction

The combination of structural steel and concrete in offshore structures is an obvious development that has finally emerged as the synergistic benefits of combining the two materials have been recognized. Two forms of steel-concrete construction offer potential advantages.

The first form is where elements of structural steel are joined to elements of concrete. Each works in its own way, but the connection must transmit structural forces in such a way to preserve its integrity. Examples of hybrid structures are:

1. Structural steel superstructures joined to concrete substructures,
2. Structural steel frames supporting exterior concrete walls and slabs,
3. Steel articulated hinge providing the articulated joint between a concrete base and a steel or concrete column of an articulated loading buoy.

For these hybrid structures, the principal concern is the working of the joint under cyclic-dynamic loads. This can properly be taken care of by prestressing the steel element to the concrete element. To ensure successful performance of such a prestressed joint, several aspects have to be given special consideration.

Because the prestress tendons are usually relatively short, the effect of takeup and seating losses will be significant. For this reason, prestressing hardware has been developed in which the final securing is done by threaded nuts or couplings rather than by wedges.

The anchorage zone of the prestressing tendons within the concrete must be carefully detailed to ensure that cracking and progressive degradation cannot occur around and behind the anchorage. The concentrated bearing of the anchorage plate produces high strains which in turn lead to radial tension: the well-known bursting forces associated with a high-capacity anchorage. For this phenomenon, transverse reinforcement is required.

Full bearing is achieved between the steel bearing plate and the concrete.

The weld must have had adequate weld procedures, and may require pre- and postheating, to ensure full development of the connection. The plate P must have proper through-thickness to prevent laminar tearing. As noted, the bearing of plate P on the concrete must be uniform. Two methods are used. In the first, the gap between plate P and the concrete is filled with cement mortar or a special bearing compound (noncorrosive) or is injected with epoxy. In the second, the concrete is ground and the plate milled, both to a fine tolerance such as 0.1 or 0.2 mm, the allowable tolerance depending on the size of the plate and its stiffness (thickness).

The second form of steel-concrete construction is where steel plates are joined with concrete so that the two form one structural element in themselves. The two must have mechanical means for developing shear between the steel and concrete components.

Typical arrangements are:

1. A steel beam is joined to the concrete by welded studs. The concrete takes compression; the steel plate takes tension and transverse shear. This is the scheme commonly employed in bridge construction.
2. Special shear connectors, in the form of transverse bars or vertical plates (with holes), are welded to the steel.

3. Two steel plates, spaced apart, are filled with concrete. The two plates are tied together with steel diaphragms, or bolts, or perhaps even short stressed bars.

The concrete ensures plate and shell action to distribute local loads and transmit horizontal shear.

The external steel plates provide tensile capacity. The internal concrete provides compressive capacity, and the diaphragms, studs, or bars provide shear capacity. With such sandwich construction, the steel plates will, of course, be forced to carry any compressive strain imparted to the concrete. Budding inward is prevented by the concrete fill; buckling outward is thus the typical mode of failure and can be restrained by the through-thickness ties. Thus not only the size of the shear connectors but their spacing and welding become important.

Where welded studs are used, the amperage and procedures must be in accordance with the manufacturer's recommendations. These welds are heavily stressed in shear and bending at points of high beam shear. The surface of the steel should be clean to develop good bond with the concrete.

Placement of concrete into sandwich composite systems requires careful planning and procedures to ensure complete filling. Small holes may be necessary in internal stiffeners and diaphragms to prevent trapping of air or bleed water. It will frequently be found practicable to leave temporary (or sometimes permanent) holes in one of the plates for access for placing the concrete by pump. The hydrostatic head of fresh (unset) concrete, especially when placed under pressure by pump, may deflect the plates.

Composite construction appears especially attractive where offshore structures must resist high local impact forces, such as those due to ice or ship and barge collision.

Composite construction is increasingly used to provide stiffness and ductility to steel tubular piling. In many marine installations, the maximum bending occurs just under the pile cap. Thus concrete fill can provide ductility, provided it is detailed to act in composite action with the steel tubular.

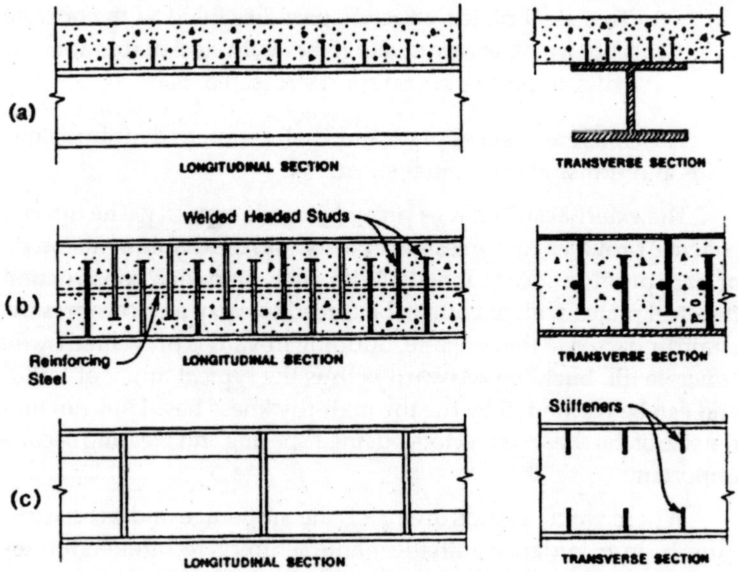

Figure 3.1: Forms of Composite Construction

The Plastic and Synthetic Materials Used in the Marine Environment

Increasingly, plastic and similar synthetic materials are being utilized in the marine environment. Glass, carbon, and aramid fibers are embedded in a resinous synthetic polymer. Uses range from glassfiber-reinforced plastic for pipelines to neoprene and natural rubber fenders and bearings, to polyethylene bags for slope protection and polyurethane foams for buoyancy. Porous geotextile filter fabrics are extensively employed under riprap to prevent leaching of sand. Epoxies are injected for repair or applied as jointing and coating compounds. Polyethylene pipes have been used for cold water pipelines in depths up to 2000 ft (600 m) off the island of Hawaii, and Kevlar, nylon, and carbon fiber mooring lines are in common use in floating offshore operations. Glass fiber-reinforced plastic is used for fender piles to protect wharves. Fiberglass and carbon tendons have been employed as prestressing tendons on an experimental basis. Ductility of concrete piles and columns has been increased by encasement in aramid fibers. Carbon fiber sheets, affixed to the bottom of beams, increase the bending capacity while carbon fiber sheets, affixed to the sides, increase the shear capacity. Aramid fibers (Kevlar) are increasingly utilized in deep-water mooring systems.

From the constructor's point of view, certain aspects must be considered.

Where the plastic is to be applied in the field, the joint surfaces must be clean, of proper texture (roughness), dry, and at the proper temperature to permit curing. Epoxies are especially sensitive to moisture and dilution by water, unless they are especially designed with a hydrophobic component to be used underwater or on damp surfaces.

Many jointing and sealing materials have a closely controlled thickness over which they are effective, with a narrow tolerance range.

Extreme temperature changes may lead to delamination of steel-neoprene bearings and fender units, due to change in properties and differential thermal contraction–expansion.

Many plastics, especially polyethylene, are subject to ultraviolet (UV) degradation, unless they are formulated with a pigment to give them added UV protection.

Most plastics are positively buoyant in water; this makes them difficult to place underwater. Some geotextile filter fabrics are now purposely manufactured to have negative buoyancy and thus facilitate placement. Fiberglass pipe may require supplemental weighting to ensure stability during placement.

Geotextile membranes and fabrics are generally supplied in rolls and sheets, which are to be overlapped in installation. The laps are a cause of many difficulties both in placement and service. Adequate, even excessive, overlap is usually a wise precaution for the constructor in order to accommodate irregularities in the seafloor and the tolerances associated with work underwater and under wave action.

Rigid plastics are susceptible to damage from impact and abrasion. They generally require special softeners for handling.

Plastic pipe such as polyethylene is subject to internal fatigue at locations of concentrated stress. Therefore, all attachment points must be reinforced and care taken to ensure a distributed rather than a concentrated bearing force. A polyethylene pipe for an OTEC project was under tow near Hawaii. Its weighted end was being held up by wire rope, which in turn was attached to the barge. Under sustained load, with cyclic stresses due to waves, the pipe ruptured, allowing the line to break loose. A major salvage effort was required for recovery.

Polysulfide sealants can be rapidly disintegrated by bacteria under some conditions. In one case in the Arabian Gulf, the disintegration took place within a few months, before the construction contract was completed and accepted.

Flexibility properties change rather dramatically with temperature, so that a material which is very pliable at 15°C may be stiff and brittle at –10°C.

Plastics are generally anisotropic in their properties, depending on the orientation and concentration of the fibers.

Plastic mooring lines, especially those of nylon, degrade over time in saltwater.

Epoxy coatings may delaminate in saltwater.

Chapter 4

The Environmental Aspects of Offshore Construction

Winds and Storms

Winds and storms are the major factors which affect the offshore construction. The predominant patterns of ocean winds are their circulation around the permanent high-pressure areas which cover the ocean, clockwise in the northern hemisphere, counterclockwise in the southern hemisphere. In the tropical and subtropical zones, the extreme heat and the interface between atmosphere and ocean create deep lows, which spawn the violent storms known as tropical cyclones in the Indian Ocean, Arabian Sea, and offshore Australia; as hurricanes in the Atlantic and South Pacific; and as typhoons in the western Pacific. The occurrence of such storms in the subtropical and temperate zones is seasonal, late summer to early fall, and is fortunately somewhat infrequent. However, while they are easily spotted by satellite as well as by observations of opportunity, prediction of their route is still highly inaccurate. Thus there may be several alerts per year at a site, necessitating adoption of storm procedures. Most of these will turn out to be false alarms, but significant delays will nevertheless occur.

Typical of the violent cyclonic storms are the typhoons of the western Pacific. These occur from May to December and have their source in the Pacific Ocean in the vicinity of the Caroline Islands. These storms travel westward, eventually dissipating over the Philippines, the China coast, and Japan or decaying into North Pacific storms. The diameters of these typhoons range from a maximum of 1000 miles to 100 miles or less. In most cases, the most severe wind, tide, and wave action occurs in a band about 50 to 100 miles wide. The storm as a whole can move in any direction, even circling back upon its previous track. It normally moves at a speed of less than 20 mph. While the predominant winds in a cyclonic storm are circular, the velocity of translation adds and subtracts. The "dangerous quarter" for navigators is that where the circular and translatory velocities join to produce a maximum.

In the Indian Ocean, east of the Indian subcontinent, tropical cyclones can occur during two seasons of the year, one the southern hemisphere cyclone season, the other the northern hemisphere cyclone season. There are many seas in which tropical cyclones were previously not thought to be present, primarily because there was so little human activity in the area. An example is the Arafura Sea, between Australia and the island of New Guinea. Not only do cyclones exist in this offshore region, but in 1982 one struck Darwin with devastating fury.

In typical offshore operations off the west coast of Australia, there will be a dozen or more cyclones spotted each season. Several, perhaps four, will be close enough to cause a cyclone alert, with all operations shut down and equipment deployed in accordance with the prearranged procedures, either towed to the safety of a harbor or, in the case of larger vessels, towed clear to ride out the storm at sea. However, usually only one, sometimes none, of these cyclones will actually hit the construction site. Nothing more serious than long swells may arrive. Thus it often is the interruption to operations rather than actual storm damage that causes additional costs and delays.

Cyclonic storms can also spin off the interface between Arctic (and Antarctic) cold air masses and the warmer air of the temperate zones; these produce the typical winter storms of the North Atlantic, Gulf of Mexico, and North Pacific, and those surrounding the

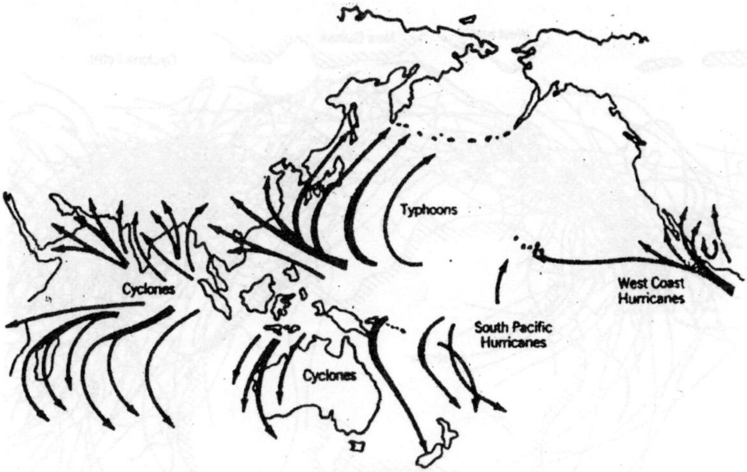

Figure 4.1: Major Cyclonic Storm Tracks in Pacific and Indian Oceans

Antarctic continent. These storms tend to act over wide areas (several hundred miles) and persist for 2 to 3 days.

As noted in earlier section patterns do exist for specific areas; for example, on the Pacific coast of the United States, 6 or 7 days of storms will usually be followed by 6 or 7 days of good weather. In the North Sea, in February, if there is a "good" day, then the probability of getting a "good" day the next day is 65 per cent. The probability for the next 2 days is 40 per cent, for the next 3 days, 5 per cent. Conversely, if today is a "bad" day, the probability of a "good" day tomorrow is only 15 per cent.

Recent activity in the Arctic Ocean has shown that during the brief summer, the interaction between the cold air mass over the ice pack and the warmer air of the adjacent landmasses can similarly spin off a series of intense but very local cyclonic storms, creating strong winds and short steep waves in the shallow open water.

Storm procedures for the constructor involve a sequence of steps: first to stop operations and secure them against the weather, second to move off site, as applicable, to a more-protected location in which to moor. Alternatively, the floating construction equipment may be moored on a survival mooring or even towed clear to ride out the

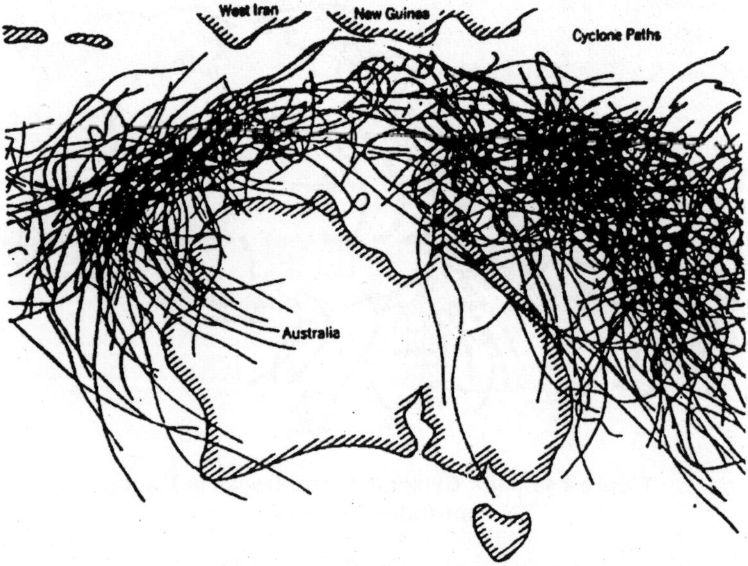

Figure 4.2: Typical Cyclone Paths Around the Australian Continent

storms at sea. These procedures will all be dealt with in later sections of this book.

It will be noted that swells normally arrive prior to the winds. A seasoned mariner will note the sky: cirrus clouds indicate a low-pressure area; they usually radiate from the storm center. The initial event of an approaching storm is often a cessation of the normal high-pressure winds, "the calm before the storm," followed by winds shifting and picking up as a fresh, intermittent breeze. It is often more dangerous to try to enter a harbor during the onset of a storm than to ride it out at sea. The harbor poses the additional problems of tidal currents and their effect on the waves, also of nearby shores and of winds that may be blowing across the channel.

Nautical terminology for wind directions is somewhat contrary. When mariners speak of a north wind, it means that the wind is coming from the north. When they speak of an onshore wind, it is blowing from sea toward land. An offshore wind blows from land toward sea. The term *lee side* is the sheltered side of a vessel away from that which the wind is blowing. However, a *lee shore* is not a sheltered shore but the shore toward which the wind is blowing and hence a dangerous shore. The windward side of a vessel is the side against which the wind is blowing.

Wind speed increases as one goes up from sea level, for example, to the deck of a platform. For instance, the wind at a height of 20 m may be 10 per cent greater than at 10 m, the usual reference height. Near sea level the friction of the waves decreases the speed significantly. Winds are, of course, not steady but blow in gusts: the 3-s gust, for example, may have one third to one half greater speed than the same storm averaged over 1 hour.

Wind storms may be of several types. There are the anticyclonic winds sweeping clockwise around the high-pressure areas over the major oceans of the northern hemisphere. Then there are the low-pressure cyclonic storms circulating counterclockwise around deep atmospheric depressions–the winter storms and the tropical cyclones.

In the tropics and subtropics, intense but very localized low-pressure zones result in sudden rain squalls which occasionally grow into waterspouts. Although high winds of short duration are associated with these phenomena, serious tornado and waterspout activity appears to be extremely rare. However, the squall winds, coming rather suddenly, can damage booms or interfere with offshore operations. Since a weather period of several days' duration may have frequent squall activity, this may require extra precautions during that season.

In many areas of the world adjacent to landmasses, high-pressure cold air masses may build up over the land, then suddenly swoop down over the sea. They have many local names, such as "williwaw"; they are associated with clear visibility, usually a cloudless sky, and hence give little advance notice prior to the sighting of whitecaps advancing across the sea. Wind velocities may reach 100 km/h (60 to 70 mph) or more. Because of their short duration, usually only a few hours, the seas are not fully developed; hence waves will be short and steep. They can catch construction operations unawares and do significant damage. Other offshore winds of longer duration are the Schmall of the Arabian and Persian Gulf and the dense air masses flowing down the slopes of Antarctica and out to sea.

Offshore winds can also blow off the desert; the infamous Santa Ana of Southern California forms as a cyclonic storm over the hot desert interiors, and its intense winds sweep out over the adjacent ocean. Because of lack of moisture, these storms are usually cloudless but full of sand and dust. Similar storms reportedly occur off the west coast of North Africa.

Milder but prevalent types of winds are the onshore-offshore winds typical of all coasts. The land heats up during the day, the air rises, and the wind blows in from the sea during the afternoon. To a lesser extent, the process occurs in the opposite direction in early morning. The onshore afternoon winds can occasionally reach 30 knots or more. These onshore winds may occur almost every afternoon on the U.S. Pacific Coast from June through August. Thus they pose a serious impediment to coastal operations.

Winds from two or more sources may be superimposed. The two most common are the high-pressure cyclonic winds plus the onshore afternoon winds. In extreme cases, these combined winds have reached 60 knots. Winds in harbors and estuaries may be considerably altered by local topography. Hills may provide relative calm in their lee, while gaps may intensify the wind speed. Similarly, winds tend to blow up- and downriver with greater intensity than over land and forests.

Storm forecasting services are available for all principal offshore operating areas. New areas, however, may suffer from lack of observation stations, especially in the southern hemisphere. Forecasting services also are not able to forecast local storms very well; all they can do is to warn when barometric pressures and air mass temperatures are right for such local storms to develop.

In the planning of offshore operations, every effort should be made to schedule critical operations for periods with low probability of storms or when those storms which may occur are of minimum intensity. Often there is a strong temptation to start early in the season or to work late just to finish up the project; these are frequently the times when storms occur and do damage that delays the project far more than a more prudent suspension of operations would entail.

It is customary in planning construction operations to (*a*) select the period during which the operation will be performed; (*b*) determine the maximum storms on a return period of at least five times the duration of work at the site (*e.g.*, a 5-year return storm for that time of year); and (*c*) develop procedures, and plans and select equipment capable of riding out such a storm without significant damage.

Sea Ice and Icebergs

Sea ice is found year round in the Arctic and during the winter and spring in the sub-Arctic oceans such as the east coast of Canada, Bering Sea, and Greenland Sea. Maximum southerly extent of sea ice generally occurs in February; the minimum extent, the retreat to the polar pack, occurs in August, leaving a circumpolar annulus of open water.

Sea ice is a unique substance, being a material that is near its melting point. It is saline, thus freezing at about -2°C. It occurs in many forms. (Reportedly, the Inuit language contains more than 20 different words to describe the different types of sea ice.)

Frazil ice forms within supercooled water, especially when that water comes in contact with foreign bodies that enable amorphous ice to congeal. Thus it can clog intakes. It also increases the frictional effect on the hull of a vessel. It may act to reduce the impact force from ice floes and icebergs.

Sheet ice is the horizontal layer of sea ice that forms on relatively calm water, freezing from the top down.

Leads are formed when the thermal contraction of the ice causes ruptures in the sheet ice. Leads may also be formed by currents and wind.

Rafting occurs when one sheet of ice is driven up over the top of another.

Ridges (pressure ridges) are formed by a combination of refreezing of water in an open lead, then closure of the lead with rafting and crushing.

A *compressive ridge* is formed when one sheet is forced normal to the face of the other, that is normal to the lead. Rafting and crushing occur, forming randomly oriented blocks of ice. Such a ridge may be from 100 to 500 m in length, but will average about 200 m long. Ridges often follow each other in closely spaced increments of 100 to 200 m..

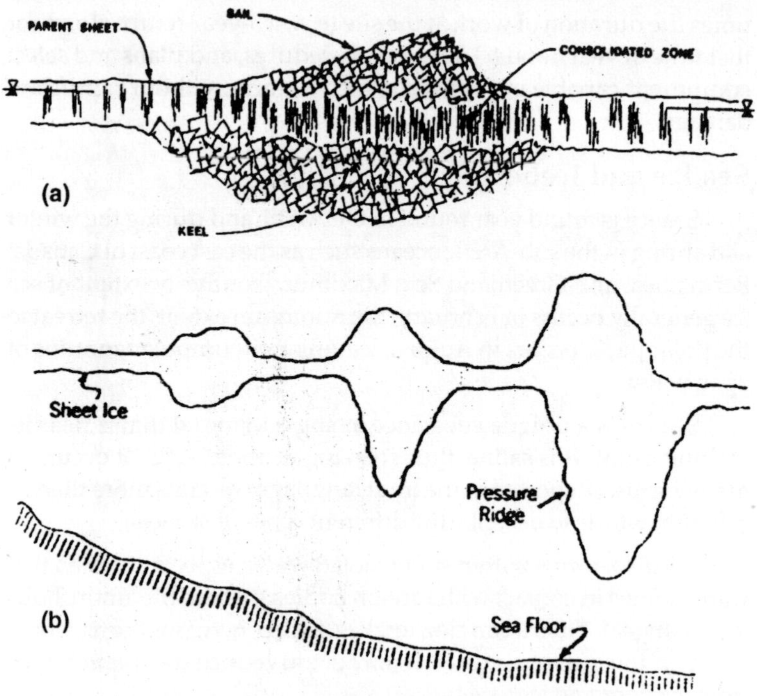

Figure 4.3: (a) First Year Pressure Ridge; (b) Typical Pressure Ridge and Sheet Ice Profile

A *shear ridge* is formed when two sheets interact laterally along the lead, crushing up a more rectangular pile of lesser size.

Ice strength is influenced by the crystal orientation, salinity, temperature, and the rate of load application.

Rubble piles are found when sheet ice pushes up against grounded ridges. These rubble piles may reach 15 m in height and be quite extensive. They consist largely of broken blocks of ice. When storm surges occur during the subsequent summer, these large rubble piles, by now partially consolidated, float off, where they become *"floe bergs,"* presenting many of the same hazards to navigation as small icebergs. Floe bergs are especially found in the northern Bering Sea. Rubble piles in the high Arctic may be formed by sheets being driven against ridges in a compressive field, that is, a region in which the ice pack is exerting strong forces on the sheet and its ridges, a large rubble pile or series of rubble piles may be formed. These are known as a *hummock* or *hummock field.*

The *polar pack* itself consists primarily of multiyear ice. As it rotates slowly clockwise around the pole, it impinges against the shallow-water annual ice which forms each winter. This annual ice is relatively stationary and is called *fast ice* or *shorefast* ice. At the boundary between the fast ice and the polar pack, a *shear zone* is set up which is highly dynamic, with much ridging and rubbing. The *shear zone* or *stahmuki* zone is usually located in 20 to 50m of water depth, which unfortunately is also a region of high prospective activity related to the development of petroleum resources.

Large blocks of polar pack ice with embedded ridges may break off from the pack, becoming *floes.* Multiyear floes may be very large, several thousand meters in diameter, but usually are only 100 to 300 m in size. Their masses are very large. Since such floes may move with the general circulatory pattern, they may achieve velocities of 0.5 to 1 m/s.

The Arctic year can be divided into four seasons: the winter when ice covers the entire sea, the summer open-water season, the breakup, and the freezeup. Within 20 to 30 km of shore, depending on the water depth, the winter ice is fast ice and hence can be used for access and transport. During the breakup, which may occur in June and July, the sea is filled with remnant ice floes and bits of ice. The percentage of ice coverage is then given either in tenths or octas, an octa being one eighth. Maximum local pressure is usually about 4 MPa (600 psi), which is a value which is often used for design of barge and workboat hulls.

The summer open water varies with location and distance offshore. It may extend 300 km at times, but elsewhere may be only 20 km. Open water may last from as little as 11 days to 90 days or so. Some years there is no open water at critical locations; this occurred in 1975 at Point Barrow.

The fall freezeup consists of thin surface ice forming, which initially is relatively easily broken up by icebreakers. Thus Canmar Drilling Company of Canada has been able to extend its working season for floating drilling and construction equipment in the Canadian Beaufort Sea to November by the use of icebreakers.

During the winter season, when the fast ice has reached a thickness of 2 m, ice roads and drilling islands can be built, enabling hauling of sand, gravel, and equipment. Airstrips can also be built on the fast ice. Dikes are built of snow and ice, and water is pumped up from holes in the ice and sprayed or flooded to thicken the ice. In the fast ice areas between the Arctic islands, insulating blankets of polyurethane have been placed on the early ice sheet, then water pumped on top to progressively freeze. During early summer breakup, wave erosion undermines the edges and thermal fractures cause large shear cracks and breakoff.

Early in the summer, when the rivers thaw and the snow in the mountains melts, the freshwater floods down over the shore fast ice. Flooded areas may be many miles in extent. Eventually the water thaws a hole in the ice sheet and large quantities of water pour through, eroding a cone in the seafloor to a depth of 7 to 10 m. This phenomenon is call *strudl-scour.*

During the open-water season, floating equipment may be employed. Some ice fragments will still be encountered and present the hazard of holing of the hull of vessels. These ice fragments may have local "hard spots" for which their crushing strength exceeds the average values by a factor of two or more. Measurements on icebreakers indicate that the usual maximum local pressure is 4 MPa (600 psi), although, recently, extreme values of 6 MPa (800 psi) have been recorded. During this summer open-water season, one or more pack ice invasions may occur with multiyear ice floes moving at speeds of 0.5 to 1.0 m/s. At such times, all floating equipment must be towed to protected water for protection.

The Arctic pack shields the open water from storms. The fetch in the north-south direction is very limited. However, the fetch in the east-west direction can be several hundred kilometers. Thus the short violent cyclonic storms that spin off of the ice pack can kick up considerable seas, with an H_s of 2 to 3 m. Such storms are accompanied by storm surges of +2 to -1 m.

In sub-Arctic areas the ice duration is much shorter. The ice itself is less strong, due to the warmer sea temperatures. However, some areas may still experience severe sea ice problems. One such is Norton Sound, where during the winter north winds break away huge sheets of recently frozen ice from the area south of the Seward Peninsula to float southward in the Bering Sea. New ice then forms and is in turn driven southward, rafting on the previous sheets.

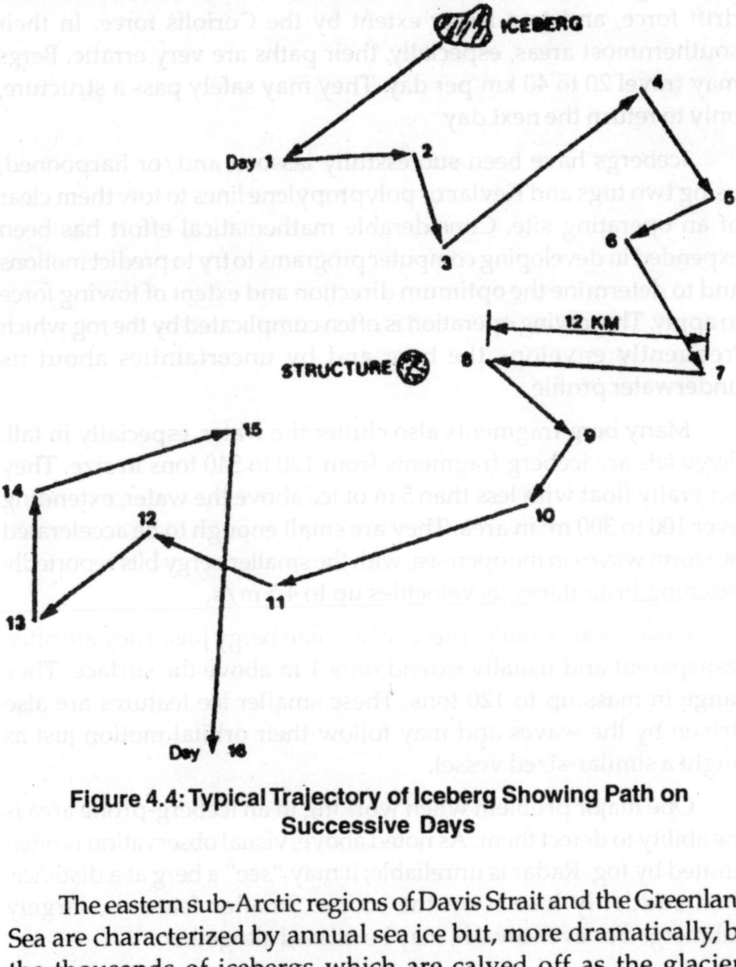

Figure 4.4: Typical Trajectory of Iceberg Showing Path on Successive Days

The eastern sub-Arctic regions of Davis Strait and the Greenland Sea are characterized by annual sea ice but, more dramatically, by the thousands of icebergs which are calved off as the glaciers discharge into the sea. These are *blocky bergs.* As they melt, they assume the dramatic shapes often seen in pictures, with pinnacles and saddles. Such icebergs may range up to 10 million tons; however, the normal maxima are about 1 million tons. About 70 per cent of the mass of the berg lies below water and may extend out beyond the visible above-water portion, a phenomenon which led to the loss of the Titanic. Bergs can also roll over whenever their centers of gravity and buoyancy shift, resulting in a loss of stability.

Bergs are driven by a combination of wind, current, and wave drift force, and to a lesser extent by the Coriolis force. In their southernmost areas, especially, their paths are very erratic. Bergs may travel 20 to 40 km per day. They may safely pass a structure, only to return the next day.

Icebergs have been successfully lassoed and/or harpooned, using two tugs and Kevlar or polypropylene lines to tow them clear of an operating site. Considerable mathematical effort has been expended in developing computer programs to try to predict motions and to determine the optimum direction and extent of towing force to apply. The towing operation is often complicated by the fog which frequently envelops the berg and by uncertainties about its underwater profile.

Many berg fragments also clutter the water, especially in fall. *Bergy bits* are iceberg fragments from 120 to 540 tons in size. They generally float with less than 5 m of ice above the water, extending over 100 to 300 m^2 in area. They are small enough to be accelerated by storm waves in the open sea, with the smaller bergy bits reportedly reaching instantaneous velocities up to 4.5 m/s.

Growlers are smaller pieces of ice than bergy bits. They are often transparent and usually extend only 1 m above the surface. They range in mass up to 120 tons. These smaller ice features are also driven by the waves and may follow their orbital motion just as might a similar-sized vessel.

One major problem when working in an iceberg-prone area is the ability to detect them. As noted above, visual observation is often limited by fog. Radar is unreliable; it may "see" a berg at a distance, then suddenly lose it on the scope. Smaller features, largely submerged or even awash, may be missed altogether.

Tides

Tides result from the gravitational pull of the moon and the sun. Due to the relative masses and distances, the sun exerts only half the influence on the tides as the moon. During new and full moons, when the sun, the earth, and the moon are in approximate line, the highest tide ranges occur; these are called spring tides. When the sun and moon are approximately 90° apart, that is, at the first and third quarter of the moon, the ranges are lower; these are called neap tides.

The depth of the sea as shown on charts usually refers to MLLW (mean lower low water), which is the average of the low-water elevations during the spring tides. Some authorities use LAT (lowest astronomical tide) as the reference datum. Since land elevations usually use MSL (mean sea level) as the zero datum, structures in harbors and rivers may be keyed to that. This often leads to confusion: it is important to verify the reference datum.

Since the lunar month is l day shorter than the solar month, the times of tidal events are constantly changing. Normally, the tidal cycle moves back (later by clock time) by about 50 min each day. The time of high tide will be 50 min later tomorrow than it is today.

Typically, there are two tidal cycles each day. One of these tidal cycles will usually have significantly greater range (higher high tide, lower low tide) than the other. Some areas in the South Pacific (*e.g.*, the Philippines and West Irian) have a prolonged high tide once each day, followed by a low tide 12 hours later. These tidal cycles appear to follow the sun; hence the peaks occur about the same time each day; with high tide shortly after noon, low tide after midnight.

The times of tide and their height for the reference station are tabulated and published 1 or more years in advance. However, the time and height of tidal peaks at a particular location depend not only on astronomical conditions but also on the local bathymetry. Tidal tables are published for most coastal areas of the world, showing the time differences and height differences from the reference station for each locality. The influence of the ocean tides takes longer to reach to the head of estuaries and bays.

The tidal range in the deep ocean is relatively minor, usually less than 1 m. However, as one approaches the continental coasts, tidal ranges may increase radically, even though the site is still many kilometers offshore. This is especially noticeable off the west coast of North Africa and in the Bay of Biscay and on the northwest shelf of Australia. Within partially enclosed bays and estuaries, the tidal range builds up, reaching its maximum near the head of the bay.

Tidal ranges vary from approximately 0.5 m in some areas to as much as 10 m in others. Here are some typical values (approximate):

Location	*Range (m)*
Boston, MA	2
Jacksonville, FL	1
San Francisco, CA	2.5
Puget Sound, WA	5
Balboa, Panama (Pacific side)	4
Cristobal, Panama (Atlantic side)	0.2
Iceland	4
Mariana Islands	0.6
Cook Inlet, AK	10
Prudhoe Bay, AK	0.5

It can thus be seen that the tidal ranges vary significantly depending on location. In the cases of extreme tidal ranges, such as Cook Inlet and the Bay of Fundy, the water may come in so fast that it forms a virtual wall of water, continuously breaking over its leading edge. This is known as a tidal bore.

The flood stage is that with significant flow on a rising tide; the ebb is the flow associated with a falling tide. Periods of little or no flow are called "slack water." The tirnes of slack water do not coincide exactly with the peaks of high and low water, since the water continues to flow for some period after a peak has been reached. Tidal currents are often stratified, with the current at the surface in a different direction from that at some depth.

Storm surges are changes in the level of the sea which are superimposed upon the tide. They are caused primarily by the effect of the wind blowing for a long period in the same direction. A secondary cause is that of different barometric pressure. When low pressure exists at a site, the water level will rise to balance. Storm surges can be 1 to 3 m in height. They can also be negative, with a combination of offshore winds and high pressure depressing the surface of the sea.

Tidal currents within harbors are greatly modified by the topography and bathymetry. They often channel themselves into separate bodies of water, each moving at different speeds, producing a tidal rip at their interface. At the mouths of rivers, the current flow is amplified by a low tide and, conversely, may be reduced or reversed by a high tide.

The other factors which affect offshore construction are rain, snow, fog, earthquakes, tsunamis etc.

Rain, snow, and fog are primarily a hazard to offshore operations because of their limitations on visibility. Fortunately, with the advent of radar, electronic position-finding, global positioning system (GPS), and other sophisticated instrumentation, these no longer constitute as serious a constraint as in the past.

Rain can occur during general storms and also during the tropical rain squall, which moves through quickly and intensely. Appreciable amounts of water can enter through open hatches or other openings in the deck; if unattended over a period of time, they can adversely affect stability due to the free-surface effect. Proper drainage must be provided, adequate to remove the rainfall at its maximum rate.

Fog is of two types. The summer-type fog occurs when warm air passes over a colder ocean. Moisture condenses to form low stratus clouds. Usually there is clear visibility at the water surface. The second type is the winter fog, with cold air over warmer water. Here the fog is formed at the surface. Yet 15 or 20 m in the air it may be bright sunshine. Rain, fog, and snow will affect helicopter operation, which requires visual observation for landing. Where fogs are prevalent, as, for example, the winter type off the east coast of Canada, then the helideck should be elevated as high as possible. In the Arctic and sub-Arctic, there is almost always dense low fog at the ice edge, where the cold air above the ice condenses the warmer moisture evaporating from the open sea.

Snow presents the additional problem of removal; else it will accumulate and freeze. When the air temperature is at or above freezing, saltwater can be used to wash the snow away, melting it as well as removing it by jet action. When the air temperature is more than a few degrees below freezing, mechanical means of snow removal may be necessary.

"Whiteout" is an atmospheric condition which occurs in the Arctic, in which the entire environment becomes white: sea, ice surface, land surface, and atmosphere alike. All perceptions of distance and perspective are lost and vertigo may occur. Obviously, this is a serious problem for helicopter operation.

Spray is created by a combination of waves and wind. Waves breaking against a vessel or a structure hurl the spray into the air where it is accelerated by the wind. A great deal of water per hour can thus be dumped onto a structure or vessel. Drains are often inadequate, having been sized for rainfall, not spray. In the Arctic and sub-Arctic, drains may become plugged by freezing.

On the Tarsiut Island caisson-retained island, an exploratory drilling structure in the Beaufort Sea north of Canada, the wave energy was concentrated against the caissons by the submerged berm. Waves reflecting from the vertical wall formed a clapotis peak (standing wave) which the wind picked up as spray. Also, waves running around the island were concentrated at discontinuities into a vertical plume, which the wind then hurled onto the island. Several hundred tons of water were blown onto the sand-filled island during a 6-hour storm. The resulting runoff eroded the surface. Tanks were damaged by the impact and equipment operation was impaired. The spray plumes reached almost 30 m into the air, endangering helicopters engaged in evacuation.

Earthquakes, although a significant load condition for the design of marine structures, are not normally a consideration during construction because of their infrequent occurrence. The most serious earthquakes occur on active tectonic plate margins such as the ring which surrounds the Pacific. These vary in intensity up to such cataclysmic events as the Alaskan earthquake of 1964. The ground motions of these large earthquakes can cause significant structural response up to 200 km.

Earthquakes also occur within plates; these typically have a long return period, so that the area may have been popularly considered to be free from earthquakes. Their magnitude is usually limited to force 6 (very rarely force 7) on the Richter scale. But their effects spread for very long distances. Finally, local earthquakes of moderate intensity (force 4 to 5) are associated with volcanism, both on adjacent islands and subsea.

Earthquake accelerations are modified with distance, the higher frequency components being filtered out, so that only long-period energy reaches the peripheral areas. Many of the structures built offshore, especially those in deeper water, and on weaker soils have long natural periods (2 to 4 s). However, the amount of energy in this frequency range is usually well below the peak energy in an active-margin-type earthquake. In evaluating the effect of earthquakes on structures, the designer will consider added mass effects and the nonlinear soil-structure interaction. Current practice for the construction of major marine facilities is to consider the earthquake with a return period appropriate to the duration of the construction, typically 20 times the duration of construction.

Although the above paragraphs would seem to apply primarily to the designer, they are of interest to the constructor because of three associated phenomena. The first is that of the seaquake, which is an intense pressure wave generated by the interaction of the seafloor and the water mass above. These short-period, high-intensity waves can be felt by ships, barges, and offshore structures for many hundreds of miles from the epicenter and usually are felt on the vessel as a shock similar to running aground or a collision. The overpressure generated is about equal to the hydrostatic pressure up to a depth of 100 m or so, then remains essentially constant below that depth.

It is now being recognized that previously unexplained cases of ship damage may have been due to these seaquakes. While not currently considered for marine operations, they may have to be considered for deep-water construction in seismically active areas in the future.

Earthquakes also generate very long period water waves, tsunamis, often mislabeled "tidal waves." The tsunami is so long and of such low height (100 to 200 mm) that it is rarely noticed on the open sea. When the energy is focused by the bathymetry and shoreline configuration, however, the kinetic energy in the wave is transformed into potential energy, resulting in disastrous lowering and raising of the water level. Such effects can spread completely across the ocean. The tsunami resulting from the Alaskan earthquake of 1964 did major damage in the Hawaiian Islands and California as well as destroying the Alaskan cities of Seward and Valdez. South Pacific events have caused tsunamis in Chile and Japan.

Certain areas are known to be focal zones for tsunamis. Harbors and estuaries are especially vulnerable. The tsunami is characterized by an initial outrush of water from the estuary, followed by an inward massive surge, the so-called tidal wave.

Chapter 5

The Marine Structures in the Arctic Ocean

Introduction

Today we are facing challenges for offshore construction in the Arctic Ocean and in the sub-Arctic seas. These frontier areas are dominated by perhaps the most severe environmental conditions yet addressed for offshore development. The Canadian and Alaskan Beaufort Sea were the primary areas of interest in the 1980s. In the 1990s the interest has shifted to the east coast of Canada, the Barents and Kara Seas, and offshore Sakhalin.

Development of the Arctic offshore areas were greatly accelerated by the leasing programs of Canada and the United States in the 1980s. Platforms subject to moving sea ice were constructed in Cook Inlet, Alaska in the 1970s and have given very satisfactory service, although occasionally subject to severe vibration from ice rafts jamming between the legs. Significant but not yet serious loss of underwater plate thickness is reported from the highly abrasive siliceous silts in the constantly moving current.

Lighthouses were constructed even earlier in the Baltic Sea, generally of concrete caissons for the substructure with steel or composite steel-concrete towers. A number of failures have occurred due to vibration-induced fatigue of steel shafts, under the cyclic crushing of the ice. Experience from studies of these lighthouses and of icebreaking vessels indicates that the crushing failure tends to become locked in resonance with the natural period of the structure.

Abrasion-erosion of the concrete has occurred in these lighthouses in a narrow band from half a meter above to 2 m below sea level. (The tidal range in the Baltic is only about 1 m). Not only has the concrete been eroded but the reinforcing steel has been torn loose. This abrasion problem has occurred only in the narrow neck of the upper Baltic where hard, low-salinity ice has moved past the structure during the spring outflow. Farther north, the hard ice has been largely fast ice; farther south, the ice has been softer, higher salinity ice, so in both these areas, there has been no significant abrasion.

The above-water decks of the concrete caisson substructures have also been deteriorated by freeze-thaw attack, especially where salt-water splash has ponded. Corrosion of the reinforcement has occurred to a moderate degree where the freeze-thaw attack and abrasion have removed or reduced the concrete cover.

The Russians constructed a prototype tidal power plant at Kislogybusk, 150 km east of Murmansk, in 1970. Although subjected to 4 or 5 months per year of primarily fast ice, there has been no reported damage or deterioration.

The Arctic Ocean proper is dominated by the polar pack, a disk of permanent ice 1500 km in diameter circulating clockwise in a gyre centered at about 80°N and 15°W. The polar pack is largely composed of multiyear ice floes interspersed with ridges and fractured by leads. The average thickness is about 4 m.

Around the periphery of the Arctic Ocean, in the shallow waters of the continental shelves, is the zone which varies from open water in the brief summer to annual ice in the winter. This annual ice is relatively immobile–hence its name, "fast ice." Between the pack and the fast ice is the highly dynamic shear zone, dominated by both annual and multiyear ridges and by massive floes which break off from the pack. This is also known as the "stahmuki zone."

Sea ice pressures show a significant scale effect, decreasing for large and global contact zones to values of 1 MPa or less, but increasing significantly to as high as 4 MPa for small local areas. A pressure of 3 MPa has been used for design of large-diameter steel tubular piles. Icebergs, being fresh water, exert higher unit pressures, up to 6 and 7 MPa.

When the moving ice encounters an obstruction, whether it be a shoal or a structure, rubble piles build up. These rubble piles may protect (buffer) the structure from further impact by floes, but, conversely, they remain as massive ice features well into the following summer, thus impeding access for re-supply.

When the ice sheet is driven onto a sloping shore, it tends to ride up onto and over the beach. When the sea level has been raised by a storm surge; the ride up may proceed 30-100 m inland from the normal shoreline. Ride-up may similarly be a threat to artificial islands and platforms.

Seasonally, the shear zone, which is where much of the potential offshore petroleum development lies, typically undergoes the following sequence:

July-September: Open water, with several intense storms. One or more invasions by multiyear ice floes.

September-November: Freeze-up. Thin ice forms, restricting movement of ice floes, but also restricting operations. Ice breakers and ice-strengthened vessels are needed.

November-May: Winter ice conditions. Movement of ice slow and erratic, driven by wind and currents. Thick ice. Rafted ice.

May-July: Spring breakup. Dynamic movement of ice, with both annual and multiyear features. Open leads. Formation of pressure ridges.

Within the Arctic polar pack are several anomalies. One is the polyna, which is a large open-water area which forms even in the winter. The other is that of ice islands and ice island fragments. These start as huge tabular bergs which break off from a glacial ice sheet on Ellesmere Island and are caught up in the polar pack. Being of glacial origin, they are freshwater ice. As they ground in shallow water, or as thermal and impact stresses cause fractures, they break up into smaller fragments.

In the sub-Arctic areas, other ice-dominated environments exist. Between Greenland and eastern Canada, in Baffin Bay, Davis Strait, and as far south as Newfoundland, icebergs are a primary consideration. These are largely blocky bergs, ranging in size up to

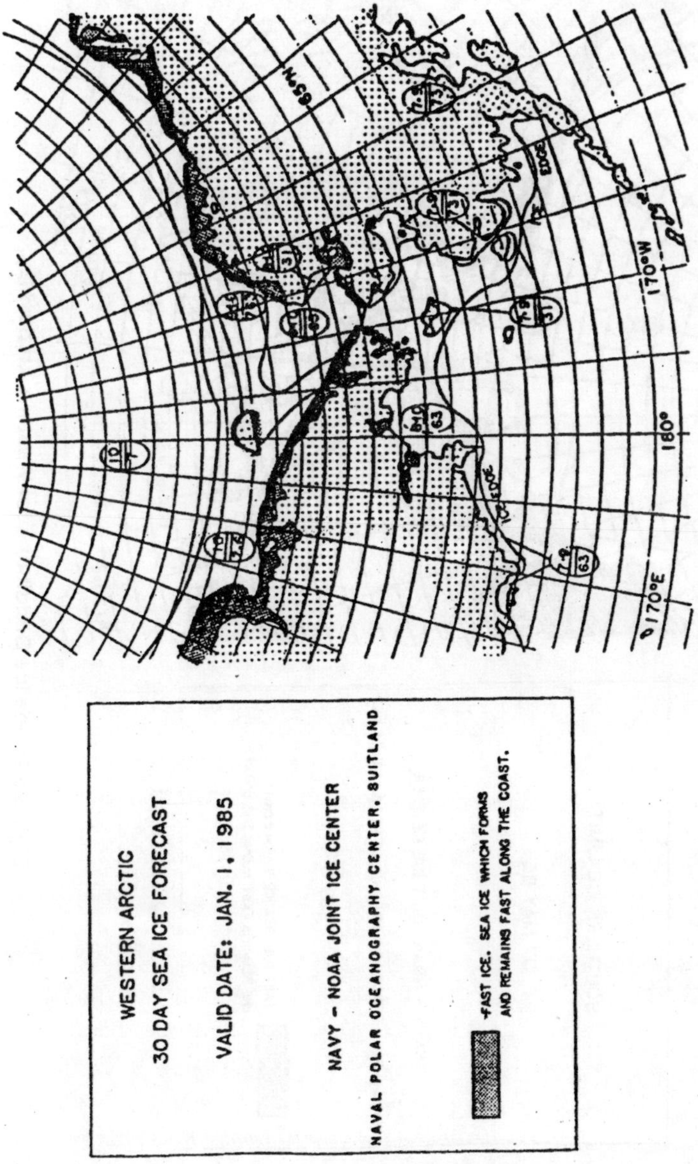

Figure 5.1: Forecast Sea Ice Conditions, Western Arctic, for January

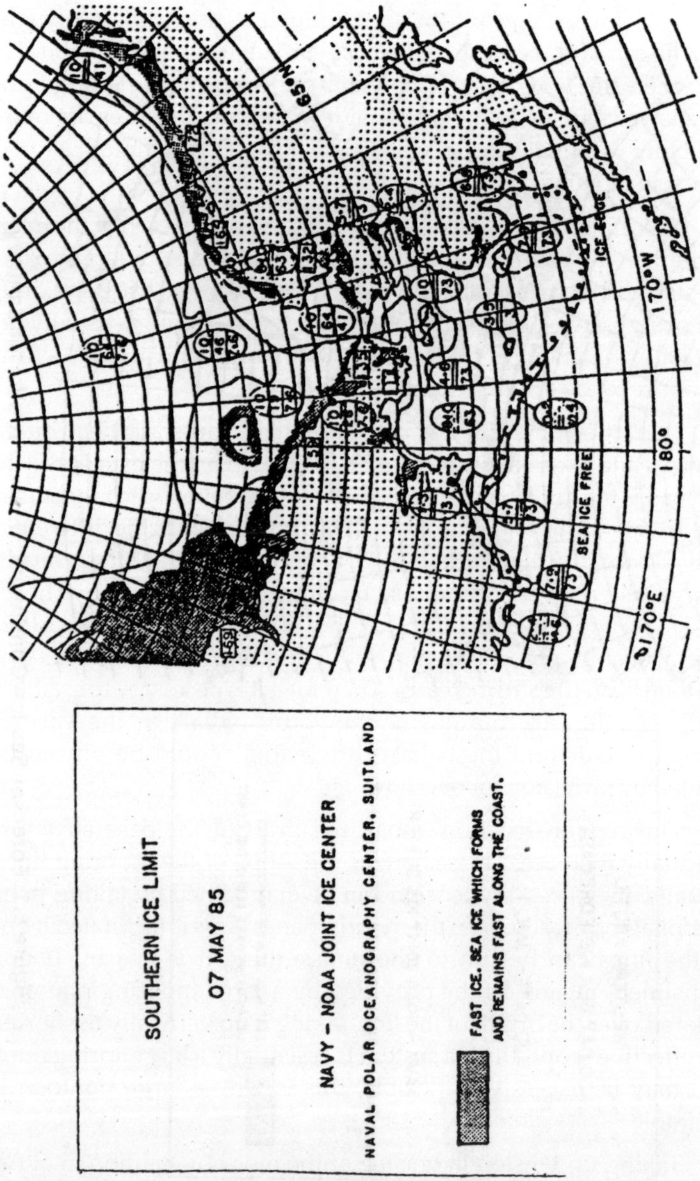

Figure 5.2: Sea ice Conditions for Western Arctic, May

10 million tons or even more. Their depth is typically limited only by the water depth. As they melt and break up, growlers of several thousand tons or so form, and finally the bergy bits, small enough to be accelerated by storm waves yet difficult to detect either visually or by radar. In the winter these bergs are locked in the annual sea ice which forms, yet movement of the agglomerated mass still continues. Other bergs ground and roll over or break up.

The worst incidence of icebergs is in April and May, which is also the worst time for fog. Annual incidence of bergs varies by almost two orders of magnitude with 100 to 2400 per year crossing latitude 48° N. Reasons for this extreme variability are not clearly understood. As noted, iceberg ice has a compressive strength of 6 to 7 MPa.

In the Chukchi Sea, the open-water season is 3 to 4 months long. The area is subject to frequent multiyear ice floe invasions, punctuated by intense storms.

The Bering Sea sees only annual ice. However, south of Nome, the ice is very dynamic, with sheet ice forming and then being driven south by the wind, leaving open water to form a new sheet. The sheets are rafted upon one another, and many annual ridges are formed. Ice driven up onto the shallow flats of the Yukon Delta forms monstrous rubble piles which last until late in the following summer. Then, in the early summer, the area is subject to occasional major storms, with high waves steepened by the shallow water and significant storm surges up to 3 m. Occasionally, the ice rubble mounds are lifted up to float off and drift northward from Norton Sound as "floe bergs." Farther south in the Bering Sea, in the Navarin Basin, the winter ice is less thick and of lower strength due to the higher water temperature. Due to the effect of storm winds, large annual ridges can still form, with consolidated zones in their mid-depth.

On the northeast sides of both Sakhalin, off the coast of Siberia, and Greenland, the sea ice forms closely spaced ridges, which constitute a compressive zone, presenting perhaps the worst ice conditions for structures and vessels. On this northeast coast of Sakhalin, southwesterly currents drive the ridges ashore and, at the same time, cause severe beach erosion.

Several atmospheric conditions have a significant effect on construction activities in the Arctic. Low temperatures are, of course, a controlling phenomenon of the Arctic, reaching to -50°C in winter. Summer temperatures are typically 10ºC. Water temperature is generally -2°C, rising in late summer to +8° near shore. Low temperatures affect not only operations but may cause brittle failure of conventional steels.

Fog tends to form at the edge of the pack ice during summer. This is due to the cold air off the ice flowing over the warm open water. Since the edge of the pack is never far offshore, this means that during the construction season, visibility can be severely restricted. Offshore eastern Canada this condition may extend well out over the open sea. This type of fog tends to hang close to the sea surface and often is limited to 10 to 15 m in height, above which visibility may be fine. An even more serious impedance to operations is that of the "whiteout," when sky, land, and sea become one vast white haze, with no reference features.

Atmospheric icing of superstructures, boat masts, rigging, and crane booms can build up rapidly and create serious problems of weight. Typically it can build to 75 to 100 mm of solid ice or 300 mm and more of porous ice. It is most serious in the southern reaches of the sub-Arctic where the air has more moisture. API Bulletin 2N warns that accretion of ice on the superstructure of vessels or structures can cause local overstressing or reduction of overall stability by increasing topside weight and the exposed areas to wind. Atmospheric icing buildup can be reduced by using tubulars instead of shapes and by means of low-friction coatings such as polyethylene.

Radio communication can be adversely affected by the Aurora Borealis ("Northern Lights") and associated electromagnetic disturbances. These are usually most severe near the time of the equinoxes (March 15 and September 15).

Magnetic compasses are essentially useless since the north magnetic pole is on Baffin Island. Loran C coverage is poor in the Arctic. However, satellite navigation is available. The Global Positioning System (GPS) gives excellent navigational control.

Windstorms are often relatively local in extent and very intense, reaching 30 m/s velocity or occasionally even more, but with a usual duration of only 6 to 12 hours. They are difficult to forecast and hence arrive with only a few hours warning or no warning at all.

When these winds occur at times of low temperature, wind chill effects can be extreme, reaching the equivalent of -70°C.

The Nature of Arctic Shore

The seafloor soils of the Arctic are among the most complex and difficult in the world. Along the shores of the Arctic proper, the shelf is very shallow for a long distance out, up to 60 km and more, at which point the depth may be only 100 m. On this shelf are recent Holocene deposits of silt (from such glacier-fed streams as the Mackenzie), degrading to clayey silts and silty clays the farther the location from the river mouth.

These weak recent sediments may be from 2 to 20 m in thickness. Near the edge of the shelf, extensive slumping has been identified, even though the slope is very flat. Below these recent sediments, in many cases overconsolidated silts are encountered up to 10m or more in thickness. These represent an extremely difficult problem for the constructor. They are so dense that it is almost impossible to drive a pile into them and extremely difficult to excavate or cut. Yet if they are broken up, the silt goes into suspension as an almost colloidal material. The most effective means of penetrating them is with a high-pressure water jet. Mechanical drills and cutters will work but experience excessive abrasion unless augmented by a jet.

These overconsolidated silts are in turn often underlain by dense to very dense sands. Subsea permafrost may be encountered, even well out into the shear zone. In the permafrost zone, the upper levels are usually partially ice-bonded, with lenses of ice, whereas at greater depths the sands are fully bonded.

Between the top of the sands and the overconsolidated silt deposits described above often lies a thin stratum, only a few meters thick, where the silt is extremely weak, almost fluid. Many explanations have been proposed for this phenomenon; one is that the water from the melting permafrost and associated gases have permeated up through the partially bonded sands to be trapped under the impermeable silts that overlie, thus breaking down the silt structure and increasing the pore pressures.

Shallow methane gas has been encountered at a number of locations. Care must be taken in drilling and pile driving to avoid accidents due to small explosions.

Occasional ice lenses have also been reported in the silt deposits, well above the body of subsea permafrost. Another anomaly is the reported presence of boulders on the surface of the seafloor, out near the edge of the shelf. These are reportedly of relatively small size (cobbles to small boulders) and hence may not be of significant concern.

The surface of the Arctic shelf, out to a depth of 50 m or so, has been repeatedly plowed by the keels of ice ridges. Furrows are generally 1 to 2 m deep, 10 to 20 m wide, with ridges forced up on the sides. Old furrows have refilled with loose sediments. The scour patterns are varied and crossed as the direction of ice movement has changed. It is postulated that the entire shelf, out to 50 or 60 m depth, has been continuously scoured over recent geologic times. Another scour phenomenon is that of strudl-scour. In the early spring, when the rivers thaw, large quantities of water flow out over the shore fast ice. Eventually they find a weak point and break through to the sea below. The velocities are high enough to scour a small crater, 20 to 40 m in diameter, 10 m or more in depth.

Pingos, the hillocks that arise on the coastal plains due to frost heave, are also occasionally encountered underwater, as relics from the period when the sea surface was lowered. Pingos can have a base diameter of 100 m and a height of 50 m. Forwardlooking sonar should be employed when navigating in Arctic shelf waters. Relic pingos have left pock marks up to 100 m in diameter and 10m deep on the floor of the North Sea and other sub-Arctic areas. These pock marks are believed to have been formed when the ice core melted, allowing the pingo to collapse.

Offshore Labrador and eastern Canada, glaciers have scoured the seafloor clean in many shallow areas. Scour marks from iceberg keels are evident at such locations as the Strait of Belle Isle. Deep scours at depths of 300 m or more have been found. These are perhaps similar to those found in the Antarctic out to a depth of 500 m, which are believed due to glacial tongues which have extended out into the sea in relatively recent geological times.

In Norton Sound, extensive sand deposits occur, often densified by wave action to depths of several meters. Overlying these sands are 1 to 2 m of very loose recent deposits from the Yukon and Kuskokwim rivers. These loose silty sands are very susceptible to movement by currents and waves. The underlying sands are often gas-charged with methane gas.

The Navarin Basin is composed of deep deposits of extremely weak silty clay deltaic deposits, susceptible to slumping even on relatively shallow slopes.

In the southern areas of the sub-Arctic, that is, offshore Newfoundland and in the North Aleutian Basin of the Bering Sea, very dense sands are encountered, the densification apparently being due to storm wave action. The sands at the location of the Hibernia platform had an internal angle of friction of 42°.

Another geotechnical phenomenon, of more concern to drillers than constructors, is that of clathrates. These methane hydrates exist as stable solids under proper conditions of temperature and pressure, but expand to 500 times greater volume as methane gas, once the limits of temperature and pressure are exceeded. Clathrates are believed to exist only at depths of several hundred meters or so. However, gas from decomposed clathrates may be one cause of the very weak silt stratum at the interface with the sands described earlier.

The sudden release of methane gas from clathrates during drilling may adversely affect foundation soils and may cause severe down drag on piles and conductors. Filling of the resultant crater may accentuate the downdrag problem. Drilling techniques have been developed to prevent this phenomenon from occurring, for example, by allowing early release of pressure before casing is installed.

The Arctic is generally an area of low seismicity. However, there are local areas of seismic activity such as near Camden Bay on the North Slope of Alaska and an active fault in the northeastern Beaufort Sea, just west of the Canadian Arctic Islands. Earthquakes in this last area may affect structures and facilities in the eastern Beaufort Sea as far south as Tuktoyaktuk, thus requiring consideration of such phenomena as liquefaction of silts and sands. Sakhalin lies between two major subduction zones and has experienced many major seismic events. The Mid-Atlantic rift runs east of the underwater plateau of northeast Greenland but at least one epicenter has been recorded in the area of potential interest.

In the southern Bering Sea, earthquakes must again be considered due to the proximity of the active plate margin just south of the Aleutian chain. For the constructor, operations in these seismic environments may involve more strict requirements for gradation of imported sands and for their densification and consolidation.

The shoreline of the Arctic roasts and barrier islands is especially sensitive to physical and thermal changes. The permafrost line is near to the surface, actually only 30 cm or so below the onshore tundra. It is this frozen soil which protects the coast, often forming a low bluff under summer wave action, with a thin layer of gravel on the beach proper. Excavations and other constructional activity can upset this equilibrium, leading to large-scale and progressive erosion of the fine sediments.

The barrier islands of the Beaufort Sea are under a more or less continuous process of erosion and deposition. Along the North Slope of Alaska, the currents are driven by the Beaufort gyre. Man-built structures, islands, seawalls, and groins interrupt this longshore process locally, leading to accretion on one side, erosion on the other.

Along with nature of shore let us also try to understand about storms which arise in oceans.

The frequent storms in the summer affect the open-water zone, which has very limited fetch in the north-south direction but much greater fetch in the east-west direction. Typically, the waves will be of only 2 m significant height, but when the ice pack has been driven northward, leaving 200 km of open water, higher waves are possible.

The wave period will be similarly affected, being usually 5 s but increasing to 6 or 7 s when the open -water fetch is extensive. Waves will tend to be quite steep due to the shallow water in the zones of most current construction interest.

Around structures and especially embankments, refraction patterns and directional spread from the rapidly moving storms may create confused and turbulent seas. The high winds pick up the top of the breaking waves as spray. The spray may become of significant amount, transporting several hundred tons of water onto a structure during a storm, creating problems of drainage. Since the spray can shoot up to 30 m or more around a structure, it can pose problems for helicopter rescue during a storm. During early fall, when the sea is still open but the air is below freezing, severe icing problems may ensue. The durations of these storms are usually relatively short, with a total duration of 24 to 36 hours and a period of peak intensity of 6 hours or less.

There are several interactive phenomena which may lead to significant storm surges, creating both raised water level (+2 m) and depressed water level (-1 m). These differences in water level are caused by the low barometric pressure, the strong winds, and the currents. The effect of surges is abnormally great in the Arctic because of the long, shallow shelf and the narrow "river" that exists for the water to escape between the land and the pack ice. This same combination leads to rather severe currents in the open-water zone, especially between Harrison Bay and Point Barrow, where the bathymetry and close presence of the ice pack lead to currents of almost constant velocity over the full water column from the surface to the seafloor.

The Bering Sea is subjected to severe cyclonic storms, not unlike those of the North Sea, especially in its southern stretches. As the storm-driven waves reach the shallow waters of the Yukon delta and Norton Sound, they are depth-limited, becoming very short and steep and piling up several meters of storm surge.

The Arctic Offshore Structures

When we study the structures which are constructed in Arctic Ocean region we can understand that all these structures have been constructed in the shallow water region. The Canadian structures have been largely built of sand, dredged and placed by hydraulic dredges when a sand source could be found nearby, or with trailing suction hopper dredges when the source was distant. In some cases, the source has been 50 to 80 km away from the site. The deepest water in which such an island has been constructed to date is 62 ft (19 m) at Issungnak, and 20 m at Mukluk, both of which involved almost 5 million cubic meters of fill.

Deposition has been largely by hydraulic or barge discharge at the surface, resulting in very flat side slopes, as flat as 15 : 1. Use of special discharge tremie pipes, with devices at the tip to slow the velocity, have resulted in denser placement and steeper slopes, about 5 : 1 to 6 : 1.

It is, of course, not only a matter of placement technique but also of gradation which determines the slope and density. Some islands have experienced progressive liquefaction when they were built up too rapidly with material that did not achieve the required density. In one case, multiple flow slides were triggered one after the other and the embankment spread out so far that it had to be abandoned.

Slope protection is required as the island emerges from the water. The sudden storms of the open-water season can quickly erode thousands and even tens of thousands of cubic meters of sand. One solution that has been adopted–with only moderate success, however–has been to build the island up to just below sea level, hold until late in the season when most of the storm activity is over, and then rapidly complete the above-water portion.

The short duration of the open-water season requires a large dredging capacity and continuous operation. Trailer suction hopper dredges with ice-strengthened hulls have been able to work up to November 1 in the eastern Beaufort Sea.

One potential problem for above-water embankments is the inclusion of ice fragments in the fill. These may thaw in the following summer or when hot oil is produced, leading to sudden slumping. This has occurred in similar construction ashore.

For the Alaskan Beaufort, gravel has been available from the river deltas of the Sagavanirktok, Kuparak, and Colville Rivers. This gravel has often been transported by ice roads on the fast ice during January, February, and March for direct dumping through the ice. Side slopes of 3 : 1 are achieved. For islands further offshore, such as Mukluk, the gravel was stockpiled in the winter and then hauled by a fleet of barges for placement the following summer. When scraped or pushed off barges, gravel has achieved slopes of about 5 : 1.

Slopes have been protected by the use of plastic bags, filled with sand, placed in one or two courses over a heavy filter fabric. The bags are UV-stabilized, 2 to 4 m^3 in size and double-strength. These have worked well for temporary exploratory islands but suffer tearing and dislocation from the ice during ice movement and breakup, requiring substantial annual maintenance.

Articulated concrete mats, attached to or laid over filter fabric, have performed well in tests and appear to be more permanent and require less maintenance. As a result, they have been adopted for use on North Star Island and Endicott Island, two of the first production islands to be built in the offshore Alaskan Arctic.

Freezing of embankments, both above and below water, has been studied by many engineers. Natural freeze-back occurs in above-water and near-surface soils. Artificial freezing has been proposed to stabilize and strengthen underwater embankments. However, there are several difficulties. First, although the natural air temperature is low, the heat of fusion is the same whether in the Arctic or in a temperate zone. Second and more seriously, there is little experience with saline soils. Such as there is indicates that brine channels are formed, with the freeze front progressively driving increasing brine concentrations ahead to form unfrozen brine lenses and hence potential failure planes. A construction dike formed around the site for the Prudhoe Bay Salt Water Treatment Plant (for the Waterflood Facility) was designed on the basis of natural freeze-back. Portions of this failed, however, apparently due to unfrozen lenses of high-salinity water.

To freeze an embankment artificially, freeze pipes can be installed from the embankment or structure in drilled or jetted-in pipes or laid as mats between layers of embankment. To be effective in increasing shear and bearing capacity for Arctic offshore structures, the temperature within the embankment should probably be lowered to at least –10°C. Frost heave during freezing must be considered. The effects can probably be minimized by following a progressive pattern to drive the expansion out from under the structure. Freezing takes time, usually several months on a large installation such as those contemplated for offshore structures. Once frozen, the condition can be rather easily maintained by either active brine circulation at low volumes or by passive means such as the thermopiles used on the Alyeska pipeline in Alaska. Frozen soils sustain short-term loads well but have high creep under sustained loads such as may occur under winter ice movements.

The soft, silty clay soils of the Arctic seafloor place severe requirements on concrete and steel gravity-base structures designed for use in the shear zone. Therefore, a great deal of attention has been devoted in recent years to the development of effective and economical methods of improving the soils. Since the silty clay and clayey silt sediments are anisotropic in character, with greater horizontal than vertical permeability, the concept of consolidation through drainage and surcharge appears attractive. To accomplish this requires the placement of an impervious membrane, and the installation of wick or sand drains. Practical considerations indicate that often the best way to carry out the consolidation is after the structure is installed, using the base of the structure plus its skirts as the membrane and draining it to the interior.

It is also considered practicable to install the vertical drains prior to arrival of the structure. An initial sand or sand and gravel embankment may be placed over the impermeable surficial soils of the seafloor. Another scheme is based on the installation of sand piles prior to arrival of the structure, designed to enhance the bearing and shear resistance of the soil initially, to be supplemented by drainage and consolidation later. Such underwater sand piles might be installed using a 1-m-diameter steel pipe mandrel with a hinged or expandable plug in its tip. This would be driven to the required depth by an impact or vibratory hammer. During driving, the mandrel would be filled with sand by means of a hopper. The top of the pipe would then be closed and low air pressure applied to keep the sand in place in the hole while the mandrel is withdrawn. Only 10 to 20 psi of air pressure is required. This method is similar to the consolidation operations at the Kansai Airport in Osaka, Japan, where multiple mandrels were mounted on a barge.

A method proposed for the Venice Storm Surge Barrier may be applicable in certain Arctic cases as well. This is the driving of closely spaced piles to consolidate the soil and to tie the strata together as a continuous block. Crushed stone is then dumped over the piles, embedding their tops. The piles are not connected to the structure except through shear. Piles may be timber, precast concrete, or steel, although displacement piles will be most effective.

Dredging of the softer materials and their subsequent backfilling with imported sand is a highly practicable and effective solution if the depth of the soft material is limited to a few meters and if the soft material will stand on reasonable side slopes during the dredging. Both these requirements are met in many areas of the eastern Beaufort Sea. The advantage of such a system is that it is positive, that it is not based on time-dependent consolidation, and that it is not affected significantly by the characteristics of the weak soils that will be removed.

Surcharging around the perimeter of a structure may be an effective means of enhancing the resistance to sliding by prolonging the shear path. This surcharge can best be applied after the structure has been installed and thus can use the structure as a reference marker for control of the dumping operations.

Another means of enhancing shear resistance is that of penetration, in which the structure is equipped with long skirts to penetrate through the soft soils. In the soft soils of the Arctic and sub-Arctic such as the Navarin Basin, the penetration may be aided by removal of the soft soils from within the skirts, for example, by jet and airlift or by jet and eductor. The eductor is essentially independent of column height, whereas the airlift is most effective with a high column.

It is only natural that early attention was given to the use of ice as a structural material for temporary structures in the Arctic. In World War I, a considerable amount of research was directed toward a mixture of ice and sawdust, known as Pi-Crete, which it was hoped would be suitable for use in constructing temporary floating airfields to enable the planes of that day to transit the Atlantic. In more recent times ice has been utilized for ice roads out over the fast ice and as temporary drilling platforms in the fast ice wedged in the channels between the Arctic Islands. The ice roads are constructed by progressive flooding, allowing each layer to freeze. Dikes are built of snow and broken ice rubble. The water is obtained from holes drilled through the ice. Such roads have carried heavy trucks and have been used for the delivery of gravel to offshore sites. The gravel is then dumped through slots cut in the ice at the site. The movement of trucks, especially when in convoy at fairly high speeds, can produce resonant undulations in the flexible floating strip and lead to failure. Spacing and speed need to be controlled to prevent this.

API Bulletin 2N, in the section "Design of Ice Roads:" states: "Attention is directed to the problems of edge loading, and of cracks opening due to tidal, thermal, wind and other forces. Thermal cracks generally appear in ice roads following construction. Such cracks are particularly noticeable following snow clearing and large changes in temperature. Wet cracks may heal- dry cracks often have to be repaired with slush or water. A particularly dangerous situation can occur if two wet cracks join to form a wedge. Attention is directed to the spacing of vehicles traveling in the same and opposing

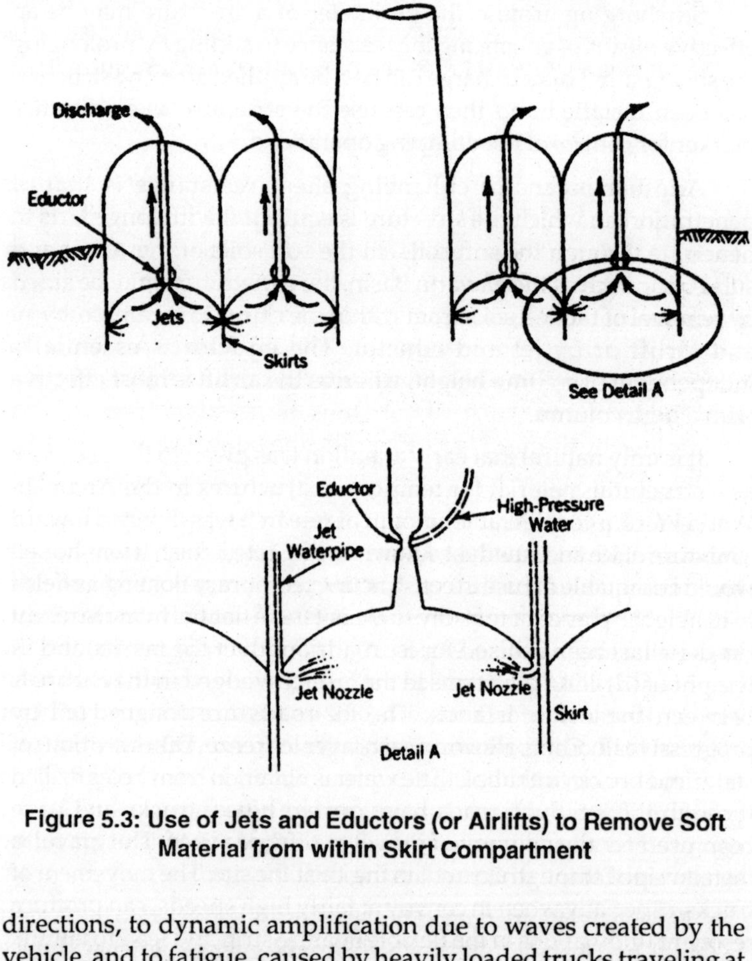

Figure 5.3: Use of Jets and Eductors (or Airlifts) to Remove Soft Material from within Skirt Compartment

directions, to dynamic amplification due to waves created by the vehicle, and to fatigue, caused by heavily loaded trucks traveling at close intervals. The life of drilling pads built up of layers of ice by progressive flooding and freezing has been extended into the early summer by the use of polyurethane insulating blankets.

Considerable experimentation has been carried out by Union Oil, Exxon, Sohio (BP Alaska), and Amoco in developing artificial ice islands for use as year-round drilling islands in the shallower waters of the Beaufort Sea. Water has been sprayed out over the ice using the large circular spray systems more usually employed in farm irrigation. This allows almost instant freezing and thus a continuous operation. A major problem with these islands occurs in spring, as the opening water thaws and erodes under the perimeter ice edges, causing the resultant overhangs to break off. Other problems encountered are propagating cracks due to thermal stresses.

The behaviour of frozen soils and embankments must be monitored and repair operations undertaken promptly. Cracks in the frozen Earth structure may be repaired by filling them with water. An adequate internal temperature monitoring system is required.

The use of ice for temporary structures will undoubtedly be extended as further work is carried out in the Arctic. Its limitations are due to the fact that ice is a solid at a temperature near to its melting point, that it, is subject to brittle fracture, and that thermal

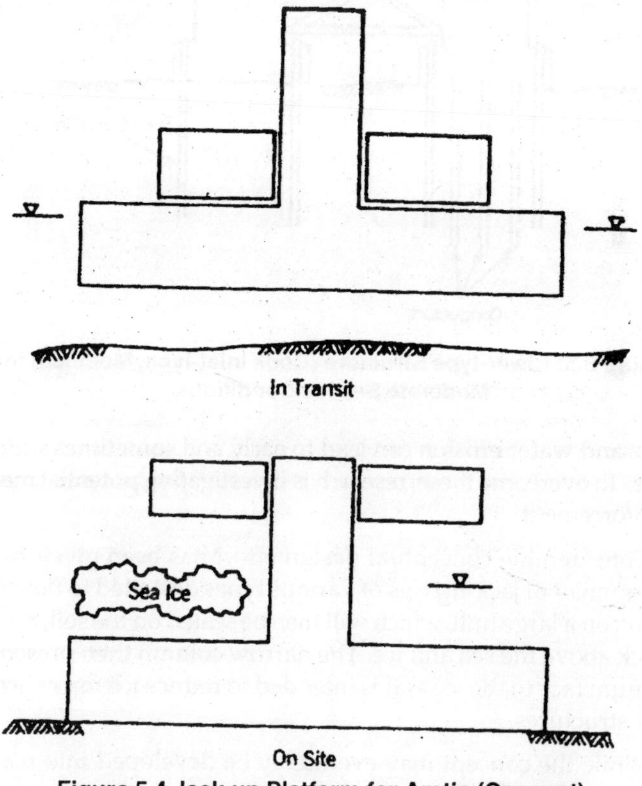

Figure 5.4 Jack-up Platform for Arctic (Concept)

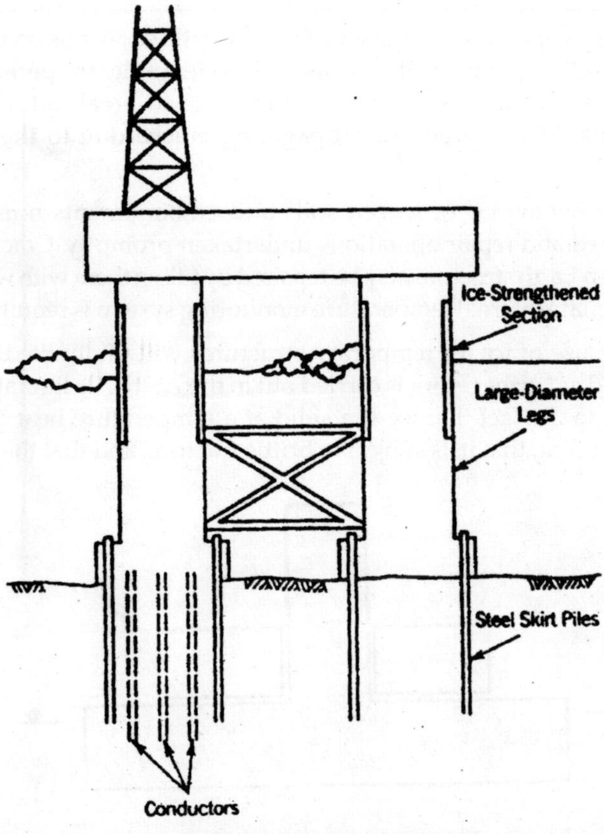

Figure 5.5: Tower-type Structure (Cook Inlet-type, Modified) for Moderate Sea Ice Conditions

strains and water erosion can lead to early and sometimes sudden failure. To overcome these, research is investigating potential means of reinforcement.

Considerable conceptual design effort has been given to the development of jack-up rigs of various types designed to float into location on a large hull, which will then be seated on the soil, raising its deck above the sea and ice. The narrow column then presents a minimum face to the ice and is intended to reduce ice forces acting on the structures.

While the concept may eventually be developed into reality, none has yet been built because of concerns over whether the extreme ice loads, such as those from a multiyear floe or short heavy ridge, arching across the legs of the structure, will really be reduced by the narrow shafts and because of potential problems of dynamic amplification under continuous ice crushing, which has led to the failure of several lighthouses in the Baltic Sea.

For deeper waters, floating structures have been developed to be permanently or temporarily moored to high-capacity anchors. These structures generally resemble the caisson structures and are built of steel, concrete, or a combination of the two. The Kulluk exploratory drilling platform, designed to work in moderate sea ice, is a floating inverted cone, so that the ice is broken and deflected downward. This structure is built of steel; an earlier design of this concept based on concrete has also been developed. The Kulluk has successfully extended its drilling season to the first part of December.

A spar-type structure in concrete was designed for the Navarin Basin in the Bering Sea, to break the ice upward or downward depending on the draft. This structure was designed for sub-Arctic areas, with unconsolidated ridges up to 10 to 15 m.

For less severe ice conditions, ship-shape hulls with turret moorings have been proposed. These are designed to weathervane with the ice movement. Thrusters may be installed to optimize this rotation and to take care of adverse combinations of wind, current, and ice. The bow will usually be wider than the body to facilitate clearance of the broken ice. Sides of the hull may be inclined inward to break the ice downward.

The major problem for all floating structures is the design of high-capacity moorings and anchors, since the global forces in heavy ice are almost an order of magnitude higher than those experienced from wave forces in such locations as the North Sea.

Once the structure is properly moored on location, a glory hole is usually excavated in the seafloor enabling well control gear to be placed below the level of potential ice scour. This has been accomplished by very powerful jets and by the use of a vertical dredge head. Glory holes are obviously well suited to floating drill vessels. An alternative to the glory hole, better suited to seafloor-founded installations, is to sink a well protector caisson (of perhaps 10-m diameter) down into the seafloor using a combination of jets and the weight of the shell along with airlift or eductor excavation. Such

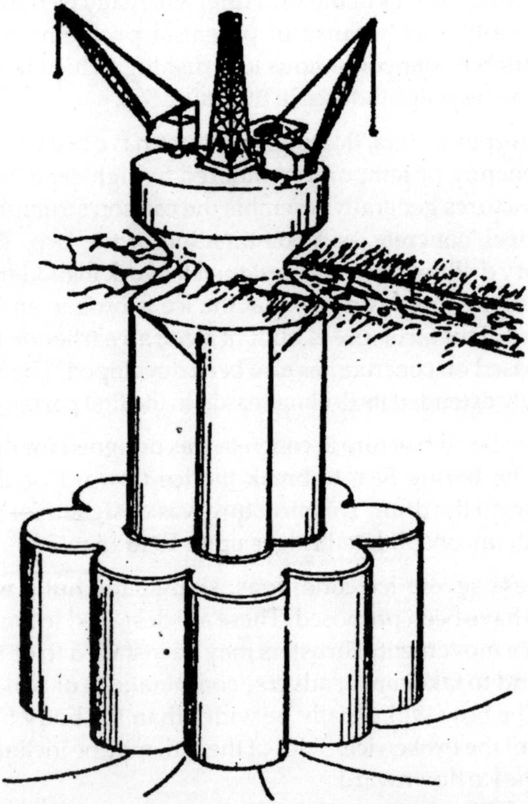

Figure 5.6: Floating Spar Platform for Use in Sub-Arctic Waters is Designed to Break Ice both Downward and Upward

well protector caissons can be either of steel or concrete. A double-steel shell can be easily transported to the site, hung under the drilling derrick, and pumped full of iron-ore slurry or concrete which will set at low temperatures. Then the weight of the fill is available to aid the penetration. Another alternative is the use of precast concrete segments, post-tensioned together vertically with prestressing bars. Off Newfoundland, large glory holes for subsea templates are excavated in the dense, partially cemented sands by vertical drills, having an enlarged drill bit (5 m) augmented by airlift. Most recently, at Terra Nova, a large diameter (5 m) drill, with air assist, has been successfully used.

The Caisson-Type Structures in the Arctic Ocean

The various caisson-type structures are typically planned for construction in the warm-water ports of the temperate regions for delivery afloat to the Arctic. While the caissons for the Tarsiut Island Caisson Retained Island were delivered on a large submersible barge, most have been and will continue to be delivered as self-floating structures, complete and monolithic, fully outfitted, to minimize the subsequent operations in the Arctic.

The tow across the North Pacific, whether from Japan, Korea, or from the Pacific Coast of North America, is so long as to have a high probability of encountering a major summer storm en route, with the relative long-period waves typical of the Pacific. It is therefore necessary to investigate thoroughly the dynamic response of the structure in such sea states. While recent advances in hydrodynamic analysis such as strip theory make it possible to analyze the response of such a structure, it is difficult to include damping aspects without a physical model test. Fortunately, in many cases these will show that the theoretical amplifications are significantly damped, so that the response of the structure remains within acceptable limits. Special attention has to be directed to cones and similar structures where the wave run-up over the sloping surface may lead to erratic lurching.

In a major storm, the tugs may have to cast the structure free. The route should therefore be selected in such a way to keep adequate sea room. After the storm the tugs can pick up the tow once again. Free-floating hawsers (Kevlar or nylon) attached to spare pendants will facilitate the pickup. Most tows will be required to be manned, in order to operate pumps if necessary. Full provision must be made for the safety of personnel men on board, in compliance with the regulations of the relevant agencies, for example, the U.S. and the Canadian Coast Guards.

Because of the length of the tow, provision may have to be made for refueling of the tugs en route. The size and number of tugs will vary with the size and displacement of the structure, but most frequently more than one tug of high horsepower (greater than 16,000 HP) will be required. Arrangements for multiple tugs will be similar to those in the North Sea. Because of the longer length of Pacific waves, towlines will probably be longer. Axisymmetrical structures such as those used in the Arctic have little directional stability, tending to yaw excessively. Skegs will usually be found effective.

Upon arrival at Point Barrow, tows will stand by awaiting the opening of a suitable channel through the ice. The depth at Point Barrow inside the shoal 7 miles offshore is a maximum of 10 to 11 m. Unfortunately, the pack ice usually hugs this shoal until September, leading to a low probability of successful passage around the seaward side of the shoal. When several class 4 to class 6 icebreakers become available, then deeper-draft tows will be able to pass north of the shoal. Their draft then will be limited only by the shallow waters of the Bering Sea and the depth of the installation site.

The towing of large offshore platform structures in the shallow waters of the Arctic under conditions of broken ice, up to 3/10 and 4/10 coverage, presents many new problems. However, it has been satisfactorily accomplished on several major wide-beam structures, the Prudhoe Bay Seawater Treatment Facility, the Kulluk, the SSDC-1, the Molikpaq, and the Super CIDS (GBS-1) being among the important achievements.

The Molikpaq and GBS-l were towed in convoy, led by a class 4 icebreaker vessel. Ice conditions became extremely heavy as the convoy approached Harrison Bay. Towing with one boat on a diagonal proved most effective. Towline snagging on ice resulted in severe impact forces on the towline, attachments, and boat. The tow speed averaged 1.3 knots over the almost 500 miles of distance from Point Barrow to Herschel Island. For the GBS-l, the tow from Point Barrow to Point Halkett took 4 days. Drafts of the two structures were 9 to 10 m.

From Point Barrow eastward through the Beaufort Sea the water depth is very limited. The underkeel clearance is minimal, so most ice must clear around the sides. If the structure is polygonal, towing with one point forward is best so that the broken ice will be forced to the side, preventing buildup of a rubble pile ahead.

Due to shallow water in the Beaufort Sea, sway, yaw, and squat are potential problems. In actual experience, low speed, the use of several boats, and the confinement of the ice have tended to minimize these adverse effects. Towing skegs may be effective in limiting the sway and yaw. The towlines for crossing the Pacific will normally have been secured below the waterline, slightly below the center of rotation, in order to keep a favorable trim, which is usually slightly down by the stern. For the Point Barrow passage, trim will need to be equalized in order to minimize draft. However, if the towlines are below water, they may tend to snag under ice blocks due to the irregular speed of the tugs in the ice. If attached above water, they may create an unfavorable trim down by the head.

On the Molikpaq tow, the initial passage was made by pulling with two icebreaker tugs. As one tug encountered heavy ice, it would slow. Then as the ice broke and the tug surged forward, the towline would snap up, causing severe impact on the fittings of both tug and tow and threatening to break the towline. In the case of the Molikpaq, they switched to using one icebreaker to break ice and another to tow; this improved the situation.

Crowley Maritime has used pusher tugs effectively to push an ice-strengthened barge to break a channel in thin ice. Consideration should be given to installing pusher notches and fittings on the stern of the structure itself and to using it as an icebreaker, since most structures are adequately designed to resist the impact of ice.

If in the future, deep-water structures will have to be towed through or deployed in pack ice, then the structure may well be configured at the towing draft waterline to efficiently break ice and to deflect the broken ice to the side so that it will not, jam the passage. Past experience (*e.g.*, the Manhattan voyage) has shown that the ice pack can be opened to pass a 30-m-wide ship. In the future, structures may be over 100 m in beam. Where skirts are installed under the structure, an air cushion could be used to reduce draft during the Point Barrow passage. Since with partial ice coverage there is a low sea state, the "water plug" could probably be reduced to about 0.5 m.

API Bulletin 2N, see. 7.3, notes: "The effects of ice accretion on mooring lines, both above and below water, should be considered. Mooring line buoyancy and drag forces, and consequently tension forces, may be affected by icing." This is an area where nylon or Kevlar towlines may be more suitable than wire lines.

Caisson structures for the Arctic will usually be installed in a manner similar to that employed in the North Sea. Upon arrival at the site, the tugs will fan out in a star formation, controlling lateral position and orientation as the structure is ballasted down to the seafloor. Positioning is controlled by a combination of satellite navigation, medium-range electronic-positioning systems such as SYDELIS, and seafloor transponders.

Some of the conical structures and monopods proposed for the Arctic encounter stability problems at this stage. There are two solutions. In waters of limited depth, a conical structure may be tipped down, keeping water plane stability until one side has touched down and then using the seafloor for stability as the structure is ballasted down to its permanent attitude. Such a system has been used in shallow waters in the Gulf of Mexico and is presumed suitable for exploratory drilling platforms in the Arctic. However, it is highly questionable for permanent production platforms because of the angle to which the deck and consequently the processing equipment must be tilted and because of potential disturbance to the foundation. The other method is to install temporary buoyant columns which give righting moment stability to the conical structure during the critical submergence phase. After installation, these temporary columns are removed.

For all gravity-base structures, as they near the seafloor, a cushion of water will be trapped underneath which must be vented internally or displaced laterally. Dowels or spuds can be dropped to engage the bottom and hold the structure in position. These spuds will have to penetrate the soil; in some cases they may have to be locked to the structure in order to use its weight to force penetration. The rate of descent can be reduced to allow the water to escape without creating piping channels. Slow descent and adequate venting of the trapped water will also help prevent lateral "skidding" of the structure.

The structure will be seated down upon a seafloor which overall is usually level, but which will have been plowed by ice scour and thus have numerous local deviations in profile and strength, with ridges and furrows, creating very high local pressures acting on the base. In order to provide a uniform foundation for a gravity-based structure, attempts have been made to preplace sandfill and then screed it off as a level base to receive the structure. This was successfully carried out underwater for the Prudhoe Bay Waterflood Project Seawater Treatment Facility for a barge-mounted facility 186 × 46 m in plan; however, this was done in the shallow water of Prudhoe Bay, within a diked-off area where the sea was essentially still.

Out in the open sea, attempts so far to screed from floating equipment have achieved only a ±300 m tolerance, despite the use of heave compensators. In the 1984 installation of Molikpaq, a tolerance of 150 mm was achieved by use of a computer-controlled drag arm from the dredge. For several of the exploratory drilling structures, such as the SSDC-1 structures, the constructors have elected underbase filling after the structure has been placed. Filling is with a sand slurry, flowed in through piping installed within the structure. The sand is first mixed into a slurry, then pumped with a very low head, approximately that of gravity, to flow out under the base of the structure. This is similar to underbase grouting as practiced under the North Sea structures, but has the advantage of being a permeable, cohesionless foundation, able to adjust to settlements, to be refilled if necessary; and to facilitate subsequent refloating.

When using sand underbase fill, the structure must first be seated, either on three or four pads that have been pre-leveled or on mud mat-type feet extending down from the structure. For the SSDC-1, the pads were concrete slabs, with a polyurethane mattress affixed, which provided support under the low bearing pressures during founding but crushed when the structure was fully ballasted, to transfer the load directly to the sand.

Skirts are generally required to seal around the perimeter of the caisson to prevent scour, enhance shear transfer to the soil, and contain sand underbase fill. Upon landing, the skirts must be forced to penetrate by the weight of the structure. In the shallow waters typical of the Arctic, the structure weight may be limited. Hence, maximum ballasting will often be required.

Chapter 6

Machinery Drives for Offshore Installations and Monitoring Systems

Introduction

Machinery drives for offshore installations range in size from fractional horsepower cabin ventilation fans to the 15 MW or more machines required for gas injection compressors.

Some concern has been expressed over the practicality of installing very large electric motors offshore owing to the stresses imposed on the structure, especially during starting. A figure of around 25 MW depending on the application (*e.g.*, less for reciprocating pump drivers) should be considered as a ceiling figure for offshore electric motor ratings.

As with onshore industrial plant, motor operating voltage levels correspond to ranges of motor power ratings, in order to keep current and voltage drop magnitudes within practical limits. For weight saving purposes, the step points at which higher operating voltages are selected tend to be lower offshore, as follows:

Voltage Level	Rating
415 to 460V	up to 150 kW
3300 to 6600 V	up to 1000 kW
11 to 13.8 kV	over 1000 kW

Higher voltages such as 22 kV could be utilized provided that the economics are favourable and that proven ranges of machine at these higher voltages are available.

Where an individual motor load represents a substantial part of the power system capacity, it is desirable to obtain designs of squirrel-cage motors, with the lowest starting current characteristics compatible with the driven equipment. Both reduced voltage starting of cage motors and the use of slip ring motors will add considerable weight to the drive package, and therefore direct-on-line (DOL) low-starting-current machines are preferred unless technical considerations for selection are overridden by those of cost and delivery.

The shaft speed of a pump, compressor or fan is critical to its performance. It is therefore necessary to obtain the required speed or speed range either by installing a gearbox between the motor and the driven equipment, or by selecting a suitable system frequency and pole configuration for the motor.

In the 1930s the US Navy changed from DC to AC systems and selected 440 V, 60 Hz as the operating parameters. Since then the whole of the NATO fleet and the majority of commercial ship designs have standardized on these parameters. This has greatly assisted in improving the availability of 60 Hz options on the standard equipment ranges of most European manufacturers. Although with small low-voltage motors the increased efficiency due to higher pump speeds at 60 Hz is marginal, with motors having ratings of the order of several megawatts the weight and power savings can be substantial.

With smaller installations, however, operating at 60 Hz may be a disadvantage where it is decided to select a reciprocating engine rather than a gas turbine main generator prime mover. The problem is that the optimum engine speed of around 1500 rpm is better suited to generating at 50 Hz. Reciprocating engines running at 1200 rpm tend to have too low a power-to-weight ratio, and operating at 1800 rpm leads to short cylinder life or even piston speeds which would be beyond the design limitations of the engine.

The higher synchronous speeds obtainable at 60 Hz also lead to higher inherent noise levels, although this can be deadened with better module insulation.

From a machinery standpoint, a major disadvantage in adopting 60 Hz for an offshore installation is related to testing the equipment prior to installation on the platform, since full load tests cannot be carried out using the British and European national supply networks. Until recently, tests were mainly carried out at 50 Hz and the results extrapolated to give projected machine characteristics at the design conditions. However, test facilities are now available in the UK for motors of up to 6 MW at 60 Hz. For larger machines, where capital investment is high and full load tests are considered essential, it is usually possible to arrange full load tests in conjunction with testing of the main generators to be installed on the offshore installation. Although this procedure is usually expensive, the costs should be more than offset by the benefits of adopting the higher frequency.

Once the system frequency has been established, it becomes increasingly expensive to change and hence may no longer be considered a variable after that point in the system design. It is therefore important to consider the number, rating and purpose of the larger drives on the installation at an early stage in the power system design, before the frequency is selected.

However, in many cases the shaft speed available from a motor, even from a 60 Hz two-pole machine (*i.e.* 3600 rpm), is lower than the required shaft speed, and it will still be necessary to install a gearbox in the drive string. Once the requirement for a gearbox has been established, changes in the drive ratio have only minor effects and the motor speed may be chosen to give the optimum motor design in terms of dimensions, weight, reliability, noise emission and so on.

The Sea Water Lift Pumps

There are three basic types of electric drive for sea water lift pumps:

(*a*) Submersible electric,

(*b*) Submersible hydraulic, and

(*c*) Electric shaft driven.

Types (*b*) and (*c*) use standard forms of electric motor as the power source; power is transmitted mechanically or hydraulically to the location of the pump. Therefore only type (*a*) will be discussed as a special application.

In floating installations or those with hollow concrete legs, conventional pumps may be used to obtain source water for cooling since the pump may be located at or near sea level. However, on steel jacket construction platforms it is necessary to draw sea water for cooling and fire fighting up to the topsides using sea water lift pumps.

Submersible pumps for this duty are of small diameter and usually very long, consisting essentially of a series of small induction motors all mounted on the same shaft. The motor-pump string, having been connected to a special flexible cable, is lowered down the suction or 'stilling' tube. Alternatively, the stilling tube may be sectionalized and the motor fixed in the lowest section of the tube. The tube is then lowered down a platform riser and jointed section by section until the motor suction is 10 metres or more below the sea surface at the height of the lowest expected tide. For cooling purposes the motor must be below the pump, and therefore difficulty is often experienced in avoiding damage to cables which have to pass between the pump and the stilling tube in order to reach the motor terminals. Bites need to be taken out of the pump retaining flanges to allow the cables past, and the cable overall diameter must allow a loose fit through these to avoid damage.

Now let us study about Diesel-electric pumps and about Downhole pumps.

Statutorily, every offshore installation has to be provided with at least two (depending on the capacity) serviceable fire pumps, each of which must be powered independently of the other. A third pump must be provided to cater for unavailability during servicing. Pumps must also be physically segregated and located geographically well away from each other so as to minimize the risk of both pump systems being damaged by the same fire or explosion. Further details on capacities of pumps etc. are obtainable from the Department of Energy's *Offshore Installations: Guidance on Firefighting Equipment*.

A typical fire pump arrangement consists of one motor driven pump powered from the installation power system, and two pumps driven by dedicated direct diesel or diesel generator sets. If a diesel generator set is to be dedicated to the supply of a particular submersible pump, weight and cost may be reduced by dispensing with any switching device and, in the manner of a diesel-electric locomotive, cabling the generator directly to the pump. The usual arrangement is to use a fire survival cable from the generator to a terminal box at the top of the stilling tube, where it is connected to the flexible stilling tube cable. The advantage of this arrangement is that the motor starts with the generator, effectively providing a reduced voltage start characteristic. A generator of lower rating than that required if the motor were started from a switching device can then be used. Another benefit of this arrangement is that a generator voltage can then be specified which minimizes the stilling tube cable diameter, *i.e.* copper and insulation cross-sections can be traded off. The fire pump diesel generator module is designed to be as independent as possible from other platform systems, and if using cooling water bled from the fire pump system itself, only requires to obtain combustion air from its surroundings.

Downhole pumps are used in the latter stages of well life to provide artificial lift when wellhead pressure or crude oil flow rates need to be improved. Downhole pumps are extremely rugged devices, having to work in a 'hole' less than 0.25 metres in diameter, hundreds of metres below the sea bed, and at pressures and temperatures near the design limits. The pump-motor string, although of very small diameter, is often in excess of 10 metres long.

The pumps are expensive proprietary devices; their manufacturers closely guard their design secrets. The installation of a downhole pump may cost in excess of £2 million; as usually no warranty is provided, that sum may need to be spent again a few days after installation if nothing happens when the start button is pressed!

However, much can be done at the surface to improve the reliability of the pump. Notably, the use of variable frequency converters to provide a very soft start for the motor has proved to be very successful in the last few years. Downhole pump motor nominal surface voltages are usually around 2kV to allow for voltage drop in the cable and to trade off copper and insulation cross-sections to minimize motor dimensions. As the pump may be capable of lifting in excess of 20000 barrels per day of oil and water from the well, the frequency converter and transformer units tend to be large.

Apart from the hazardous area requirements, the motor drivers for water injection pumps will be very similar to those used onshore. Owing to the pressures involved (typically 200-400 barg) care needs to be taken in the design of controls to ensure that no undue stresses are put on the pump or pipework, particularly during startup. It should not be possible to start the motor if the pump-motor set is running backwards owing to the wrong valves being inadvertently opened. As well as causing shock to the pipework, such maloperation may draw starting currents of excessive magnitude from the system, possibly causing damage to the motor windings and/of power system instability.

Types of Battery Used in Offshore

The following types of Battery are used in offshore.

1. Lead Acid Plante Cells
2. Lead Acid Recombination Cells
3. Nickel Cadmium Cells

Lead Acid Plante Cells

This type of cell is appropriate for most applications where a suitable ventilated battery room can be provided. It should not be used in a hot vibrating environment such as adjacent to engines. This is because the electrolyte will tend to evaporate and require frequent topping up owing to the heat, and the vibration will tend to cause the oxide coating to fall off the plates. With careful maintenance these batteries may last in excess of 20 years.

Lead Acid Recombination Cells

These cells are a relatively recent development and provide several advantages over Plante cells. The cell contains no free electrolyte, all the electrolyte being contained in an absorbent blotting-paper-like material. Except for a small safety vent, the cells are scaled for life; unlike Plante cells, they normally emit only molecular quantities of hydrogen. Because of their recent availability, no operational life expectancy figure is available, but it is known from present experience to be in excess of eight years.

Excessive charging of recombination cells will permanently damage them by electrolysing the small amount of electrolyte they contain. Once this has been vented, it cannot be replaced. Therefore it is vital that, with the chargers used, there is a very low probability that a fault will occur which allows charging voltages greater than the maximum recommended by the manufacturer. A boost charging facility must not be fitted for the same reason. Also for the same reason, engine starting batteries of this type must not be connected to an engine driven charging alternator, as this will more than likely exceed the permitted charging voltage.

If, on discharge, the cell voltage falls below a certain minimum value, it will no longer be possible to recharge the cells affected, making it necessary to replace them. Therefore it is important that during the installation and commissioning period, cells are fully recharged at least every six months.

Despite the above considerations, these batteries are becoming very popular, mainly because of the saving that can be obtained from the reduction in maintenance required, and since reliable electronic chargers are available with good voltage regulation and current limiting facilities.

Nickel Cadmium Cells

These cells are generally more able to withstand heat, shock and vibration than both of the lead acid types. A nickel cadmium battery is 35 per cent lighter than the equivalent lead acid one. However, nickel cadmium batteries are considerably more expensive, especially in comparison with recombination cell batteries of the same capacity and voltage. Although able to endure longer periods of inaction without charge, they may require more skilled maintenance than lead acid batteries for the following reasons:

1. There is a memory effect associated with repetitive cycling of the battery which may cause the battery voltage to be depressed below the value expected from its discharge characteristic. The effect is usually temporary and can be reversed by reconditioning cycles.
2. Sustained boost charging may have the effect of depressing the discharge voltage slightly. As above, this can be reversed by reconditioning cycles.

It is common to find a boost charge mode available on battery chargers, the intention being to enable the operator to reduce the battery recharge time. With the electronic chargers now available, battery recharge durations are short enough for most purposes without recourse to a manually initiated boost charge mode. Boost charge facilities are not recommended on any battery systems because of the excessive gassing they can cause. If they must be fitted, some form of timer should be incorporated in order to prevent sustained overcharging. They should not be fitted to the chargers of sealed lead acid cells if permanent cell damage is to be avoided.

Special provisions must be made for the housing of batteries offshore in order that the risks due to electrical ignition hazard, evolved hydrogen and electrolyte spillage are minimized.

The battery room provided should be lined on walls and floor with a rubber based or similar electrolyte (either sulphuric acid or potassium hydroxide) resistant material. Owing to the violence of any contact between the two forms of electrolyte, it is not advisable to house a mixture of lead acid and nickel cadmium cells in the same room. Inadvertent topping up with the wrong type of electrolyte would also be extremely hazardous for the personnel concerned.

Both types of cell evolve hydrogen, and this must not be allowed to build up to explosive levels. A ventilation system must be provided sufficient to dilute the hydrogen well below this level at all battery states. A hydrogen detector is usually provided, and any increase in hydrogen concentration would be indicated at a manned control room for the installation. Cells should be racked in tiers so that each cell is easily accessible for maintenance or replacement.

To cater for major gas leak conditions, an explosionproof isolator must be provided for each battery. These double-pole isolators would be remotely operated through the ESD system should such a serious gas leak occur.

Although very little hydrogen is emitted from lead acid recombination cells, it is still advisable for batteries of significant size to be located in a dedicated purpose-built room. Failure of the battery charger to limit charge voltage would lead to a short-lived evolution of hydrogen, and the room ventilation would need to cater for this. A dedicated room also prevents the access of unauthorized personnel who may not be aware of the dangers associated with stored electrochemical energy; tools or other metal objects inadvertently placed across battery terminals will cause heavy short-circuit currents to flow.

The Gas Detectors

Most gas detectors used offshore employ a conducting element coated with a catalyst material which causes the element to conduct a greater DC current in the presence of a hydrocarbon gas.

Although these devices include voltage regulation circuits, it is advisable to provide a DC supply well within the tolerances specified by the gas detector manufacturer, and to check battery voltages regularly, with chargers switched off, to ensure the correct voltage is available. Low voltages or sudden voltage changes may cause spurious alarms. This type of detector can become poisoned by the presence of chemicals such as paints and aerosol sprays which act in a similar manner to hydrocarbon gas, leaving the device continuously in the alarm state. A buildup of salt and grime may prevent gas reaching the gas sensitive part of the device so that it fails to an unsafe state. Such failed detectors must be found by regular testing and replaced. Infrared gas detectors have recently become available which can detect a range of gases and are impervious to poisoning. Open path types will monitor over some distance and are useful in external situations. It is necessary, however, to clean the lenses on the devices regularly.

Gas detectors are placed at strategic positions in process areas at points such as wellheads, separators and gas compressors, near flanged joints and in other places where leaks are comparatively more likely. Ventilation air intakes must all be monitored, especially those for non-hazardous areas such as accommodation where most electrical equipment will not be designed to prevent ignition, and where smoking is normally permitted. Gas detectors are not very fast acting (catalytic type more than 10 seconds, infrared type 5 to 8 seconds) and also, for reasons given above, require frequent testing and maintenance to ensure a high probability of detection should a gas leak occur. For this reason, gas detectors are normally provided in pairs or sets of three or more, with 2-out-of-N voting logic. Gas detectors are also required in the intakes of engines to detect the presence of explosive mixtures in the intake ducting and, in the case of a turbine, the turbine hood. Detection of gas in such locations should lead to immediate shutdown of the engine in order to prevent damage due to explosion or overspeed. However, ingestion of gas from process leaks by governed diesel engines is unlikely to cause an overspeed, owing to the need for compression ignition.

Combinations of different types of fire detector are used to obtain optimum cover for a particular situation. For example, a gas turbine hood will usually have a combination of heat and ultraviolet detectors. To improve reliability and to avoid spurious operation, several detectors are often operated in a voting arrangement so that, for example, one detector being activated will give an alarm whilst two or more will discharge extinguishant. As mentioned above, for gas detectors the 2-out-of-N logic is preferred.

Detector circuits are supervised for continuity and short-circuits by supervision circuits, and all monitoring logic operates on the fail-to-alarm condition principle, *i.e.* the occurrence of any failure which the monitoring systems are designed to detect will initiate an alarm. The cabling for detector circuits should be routed back to the panel so as to prevent one fire from disconnecting all the detectors in one particular area.

Chapter 7

The Importance of Offshore Construction

Introduction

The importance of offshore construction can be noted as follows:

1. Hybrid concrete-steel platforms
2. Single-point moorings
3. Articulated columns and loading platforms
4. Seafloor well templates
5. Underwater oil storage vessels
6. Cable arrays, moored buoys, and seafloor deployment

Most, but not all, of the above are employed in the development of offshore petroleum resources. However, many of these structures are also used by other industries and for other functions. Single-point moorings have been employed for slurry transfer of iron ore and coal, and tension-leg platforms for a variety of military installations.

In the actual sea environment, the lifting vessel, typically a derrick barge or drill rig, rises and fall in heave, augmented by pitch, while the large base is essentially fixed in position, unable to respond because of the huge added mass of water above, plus inertial effects.

There is, of course, some flexibility in the boom and stretch in the wire rope. Lifting over the side will permit additional softening due to roll. However, these are generally inadequate when large base caissons are being emplaced. Hence, some for of heave compensator has to be employed. Further, the main function of the derrick barge or barges is to control the rate of descent. Hence, in practice, stability should be achieved from the basic relationship of center of buoyancy to center of gravity. Large spar buoys may be used to control descent.

Free-surface can be controlled by subdividing the interior of the base so that it still has a small positive buoyancy when a number of compartments are full, the rest empty. Then small control tanks at the corners can be used to ballast to the slight negative buoyancy needed for sinking. These small compartments have minimal free-surface effects.

The requirement for some large compartments to be completely empty while others are completely full obviously places high structural demands on the intervening bulkheads. To minimize the additional structural weight, consideration should be given to tubular steel struts. Air pressurization is, of course, another means.

This unbalanced pressure in adjacent tanks must be limited by installation of some positive means to prevent the external sea pressure from inadvertently pressurizing the full compartments. A vertical standpipe, designed to withstand the pressure and founding depth, and air-filled at atmospheric pressure is one method.

As the base submerges below water, there will be hydrodynamic effects from the waves. The so-called "beach" effect will give a lift force which may necessitate adding additional ballast in the control tanks.

Once the base is founded and fully ballasted, the steel jackets or tower is installed, typically four legs will be inserted into matting sleeves. Less risk is involved if the dowels are pre-attached to the base and the sleeves are in the jacket legs; then there is no risk of the legs punching through the roof of the caisson.

The Harding gravity-base tank was placed by the used of three strings of multiple buoys. Then the steel jacket was placed on top.

Obviously, sophisticated positioning devices and instrumentation will be used to control the final jointing. ROVs with video can be employed. Separation distance can be monitored by short distance sonic transducers. Orientation can be shown by gyro compass.

Mechanical guides such as conical funnels should be used final centering. Consideration can be given to making one dowel and one sleeve combination longer than the others, to engage early. The contact should be cushioned. Hydraulic jacks can be extended prior to contact to control the impact forces.

The Advanced Single Moorings

Single-point moorings have been developed over the last several decades. The initial concepts were designed to enable tankers to moor and offload or unload crude oil. They represented an economical solution for transfer of oil wherever sea conditions permitted a reasonable usage, for example, 65 per cent availability. Single-point moorings have subsequently been improved and extended to enable them to handle several different petroleum products and even to load iron ore slurry into ore carriers.

The principal construction operations include the installation of the base structure and the anchors. The connection of hoses requires a support rig with hoisting capabilities and divers. The installation of the buoy is mainly a question of positioning, followed by the more tedious job of equalizing the tension in the several anchor legs.

Bases have been usually constructed of steel and are initially buoyant. They are sunk to the seafloor by ballasting down in calm water, with decent controlled by lines from a barge working over sheaves. The elastic stretch in the lines is counted on to absorb the dynamic loads, which are principally due to wave action on the support barge. Ballast water is, of course, confined within small compartments in the base in order to avoid free-surface effects. Bases have also been constructed of reinforced concrete and transported to the site afloat.

Bases are usually relatively small, 10 to 20 m in diameter, 3 to 5 m in height. Even though relatively small compared to the larger structures previously discussed, their installation requires consideration of dynamic lift forces and stability.

Once the upper surface is below water, there is large hydrodynamic added mass in heave. The barge tends to rise with the waves while the base structure acts as a "sea anchor." Elastic stretch in the lines, and deflection of the boom can accommodate limited differential wave. At the same time, the loss of water plane on the base makes it necessary to provide a substantial righting moment from the crane hook in order to maintain stability. This element of the righting moment is calculated as $P \times l$, where l is the height from the hook to the attachment points on the base and P is the lift.

Once on the seafloor, the base are filled with slurried sand or slurried iron ore or with the cement grout pumped through a hose. Grout-intruded aggregate has also been used, even though there may be practical difficulties in placing the aggregate and in the multiple grout injection points required. Grout-injected aggregate was successfully used on the Polaris Missile test pad of San Clemente Island, California, which were of similar size and depth and contained very complex inserts. Tremie concrete may also be employed. With cement-based grout or concrete, the effects of heat hydration must be considered.

Underbase grout may be placed through a hose to fittings on the base structure, in a manner similar to that of gravity-based structures. Since the skirt length on the base structure is necessarily limited, there being very little net weight to cause penetration, the grout should be highly thixotropic so not to escape out to the open sea. Drop curtain of canvas or sandbags may be used to seal the bottom edges or steel skirts may be provided.

Scour protection may be required. Since the water depth may be less than 30 m, the combination of currents plus wave-induced movements can cause serious erosion. Filter fabrics covered by rock are typical, but other clever schemes have been developed, including "artificial seaweed" (closely spaced nylon ropes) hanging from a ring around the structure to slow the water movements. This is intended to cause deposition instead of scour.

The base often is being fitted with a manifold on its top, with fixed curved pipes (the "PLEM") leading over the edge of the base, through which hoses are connected to the pipeline from shore.

Bases may also be pile-supported in soft soils. Piles are driven through sleeves in the base and connected by grout. These piles will be relatively short and lightly loaded; their main function is to prevent

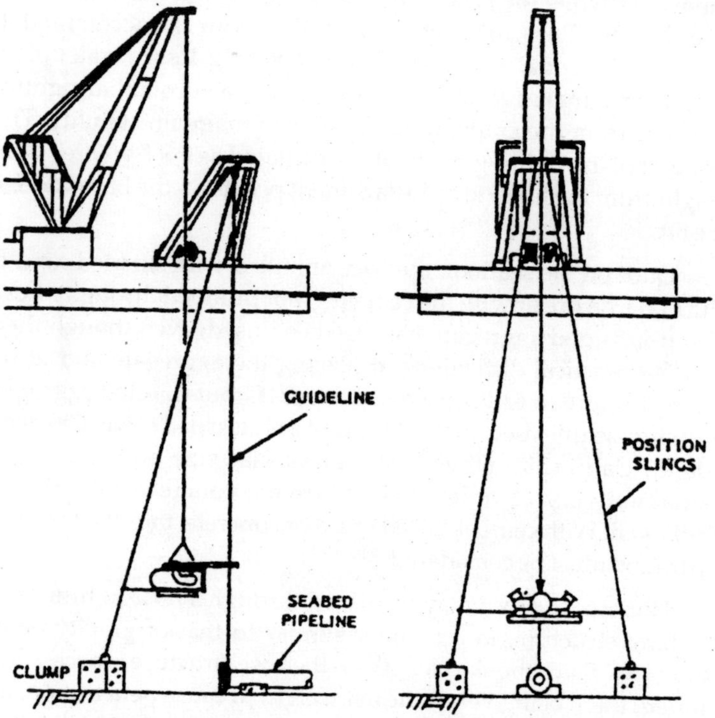

Figure 7.1: Installation of Calm Single-point Mooring for Offloading Oil : Guideline Mooring System

excessive settlement, tilting and sliding. Anchors for the several types of single-point moorings can be large gravity blocks or driven or drilled in piles. In several cases, in hard seafloors, holes have been drilled and heavy anchor chain run into the holes, after which they are filled with tremie concrete, using very small aggregate, for example, 8 mm, in order to permit placement through a flexible hose guided by a diver. A steel transition flare piece, like a funnel, is set at the top to prevent excessive wear on the chain at that point.

Single-point moorings are increasingly being used to moor tanker for storage and especially for floating production systems. In the latter case, they are combined in the field development with subsea templates. When used for semipermanent moorings, the system must, of course, be made heavier and more reliable to safely hold the vessel, even during storm conditions. While the great preponderance of current development is directed toward the use of floating production systems for marginal oil fields, the basic concept is also suitable for offshore processing storage, and transfer vessels for other commodities.

A natural extension of the more advanced single-point mooring systems has been the development of the articulated columns or articulated platforms, such as those installed in the North Sea for the offloading of crude oil. These differ from the single-point mooring (SPM) systems primarily in structural size and off loading rate.

These platforms consist of a base, a column, and a duck structure. The base is anchored to the seafloor by gravity weight or piling. The column is buoyant, and its center of buoyancy is well above its center of gravity. The base and the column are joined by an articulated hinge, called a cardan, which allows articulation on both axes but is restrained against torsion.

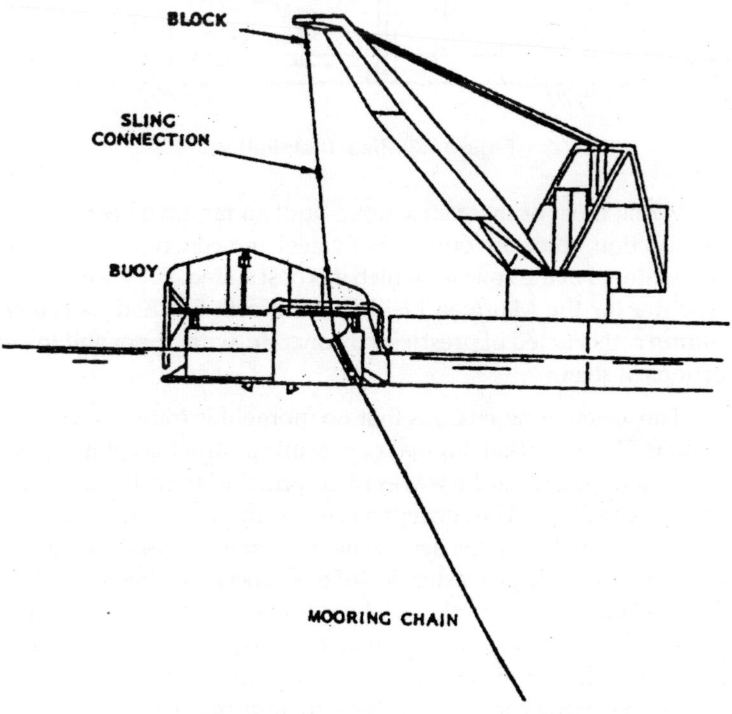

Figure 7.2: Mooring Chain Connection

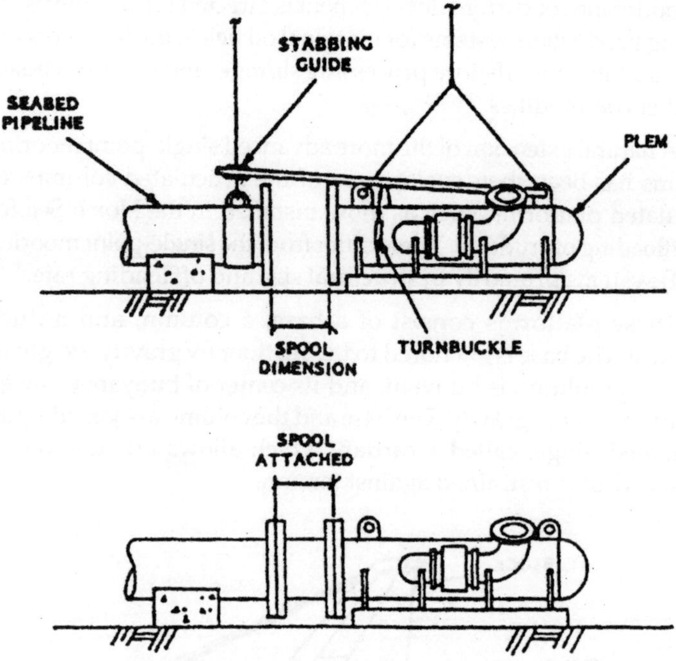

Figure 7.3: Plem Installation

While most of these structures built so far have been of steel construction, some are built of both steel and concrete, employing each material in the zone for which it is best suited. For example, the structure for the Maureen Field in the North Sea had its central column constructed of prestressed concrete, with upper and lower sections of steel.

The articulation ensures that no moment is transferred to the seafloor. The reduction this makes possible in structural dimensions consequently reduces the waves force, and the lateral force tending to produce sliding. The concept has been applied to flare stacks, off loading terminals, and the deep-water mooring of vessels for floating production systems. An extension of this concept has been proposed by C.G. Doris, for example, for deep-water (500 to 800 m) drilling and production platforms. The installation of these systems can be quite complicated and requires the sophisticated application of advanced hydrodynamics and construction engineering.

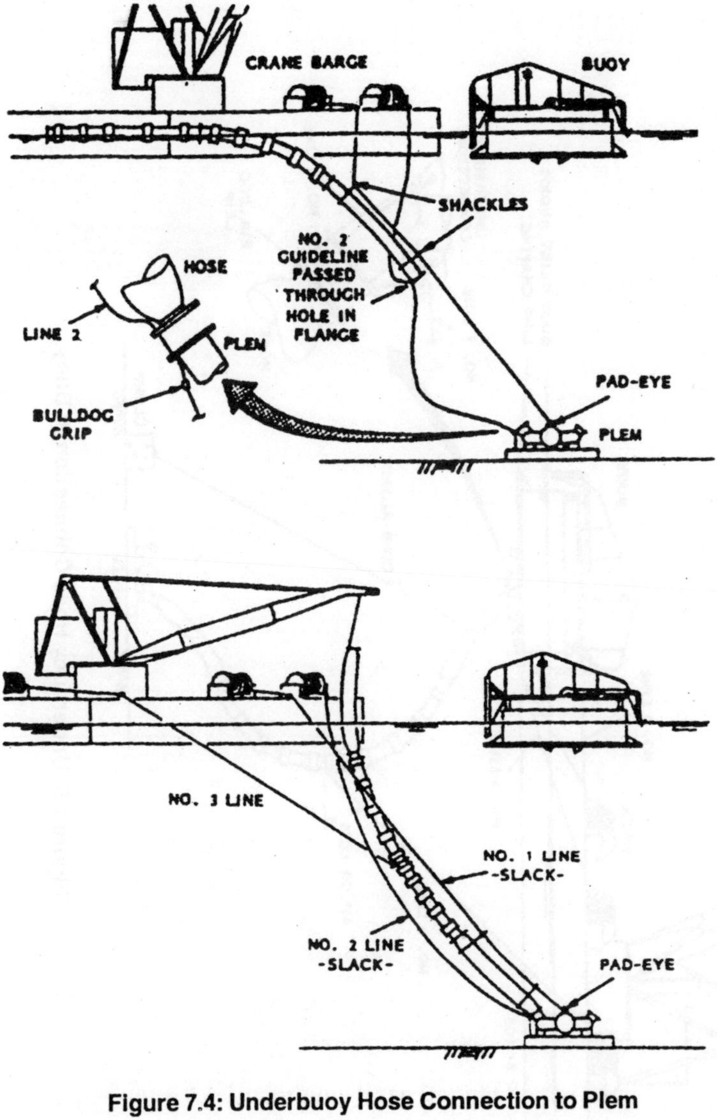

Figure 7.4: Underbuoy Hose Connection to Plem

An initial construction concept was to install the base first, ballasting a steel or concrete cellular structure to slight negative buoyancy. As has been seen with other fully submerged structures, the dynamic response during lowering of a moderately large base is very significant, especially when coupled with an offshore derrick

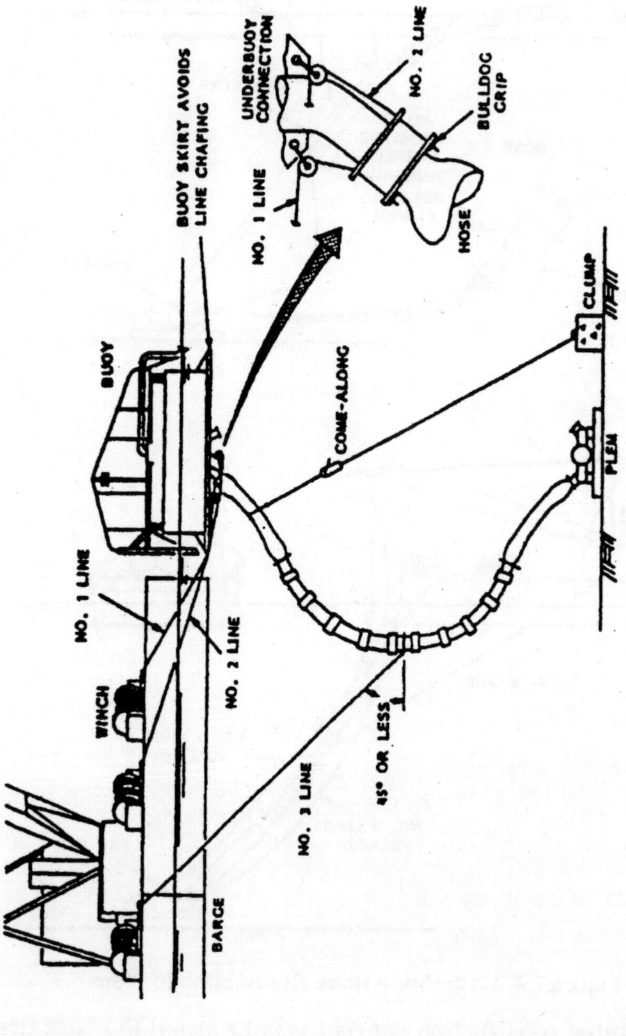

Figure 7.5: Underbuoy Hose Connection to Buoy

barge which is being accelerated in heave, roll, and pitch by the waves. Since these bases are much smaller than those required for support of a platform, the use of a spar buoy as a heave compensator is sometimes adequate. The lowering line will then be horizontal at the surface and be controlled by a winch or the barge or even by a tug.

Once a base is seated, the column is brought to the site floating horizontally, upended by ballasting to the vertical attitude, and then guided and ballasted down to a mating of the hinge. Diver intervention may be required at the mating to drop in the securing pins and activate the hydraulic locks. Tensioned guidelines have been used to control the mating. Often the final connection is made by pulling line which directs and pulls the mating cone into position.

Another method of lowering the base has been to use a "controlled free-fall" system in which buoys are attached at intervals to lines leading to the barge or boats. As the base structure is progressively ballasted, it descends until restrained by the buoyancy of the next set of buoys. Then more ballast is added, either to the base structure or to one of the buoys, so that the base may descend another step. In this way, velocity of descent is kept low, dynamic forces are accommodated, and control is maintained, even to the extent that the process is reversible by ejecting the water from a buoy by compressed air. However, this conceptually attractive system has encountered some practical problems in actual application. When the prototype underwater oil storage "Seatank" was being installed by this method, the lines joining the buoys become fouled due to waves and wind. Waves washing over the shallowly submerged tank created dynamic instability, leading to eventual total loss of control. The structure plunged to the bottom and imploded.

As water depths have increased and structures have become larger with greatly increased functional requirements, other methods have been developed in order to give proper control and reduce the amount and complexity of underwater operations, especially that of mating the critical cardan joint.

One such method is to attach the column while the base is floating at the surface. This enables the subsequent upending and decent to be fully controlled with regard to stability and depth. It also ensures that the cardan assembly is carried out properly and facilitates inspection. The base raft is titled by ballasting so that the socket for the cardan is just above water. The column, floating horizontally, is ballasted to raise the cardan above water. Now the mating can be carried out in the dry. Control of orientation is, of course, one of the major problems in the initial assembly. Since it can be carried out in shallow, protected water, lines to mooring buoys and to anchors can be employed.

Once the articulated column is mated, it can then be towed to the site in either the horizontal or vertical mode. Smaller structures, with minimal facilities, such as mooring structures for floating production systems, are usually towed in the horizontal mode, whereas structures with extensive topside facilities will be upended to the vertical while still in the harbor, the topsides installed and hooked up, and the structure towed to the site in the vertical mode, to be ballasted down onto the seafloor. To upend the combined base-column structure, iron ore ballast may be slurried into the base.

One of the more demanding of such installations was that of the Beryl A flare stack, where the articulated structure had to be not only upended and seated properly, but had to have an exact distance and orientation from the production platform in order to enable the prefabricated flare bridge to be set. Lines were therefore run from the

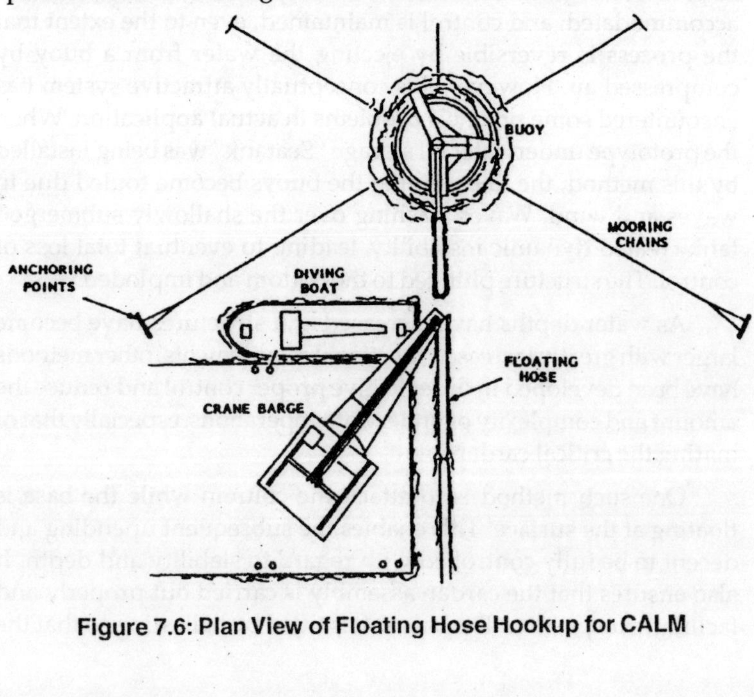

Figure 7.6: Plan View of Floating Hose Hookup for CALM

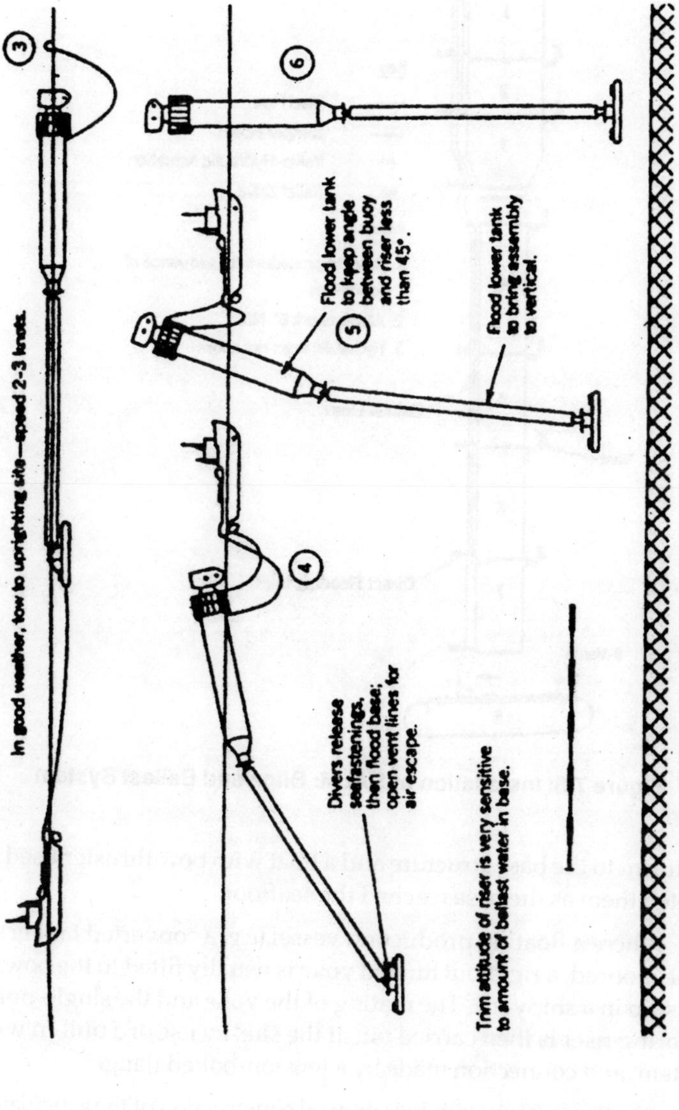

Figure 7.7: Installation of SALM: Steps 3 through 6

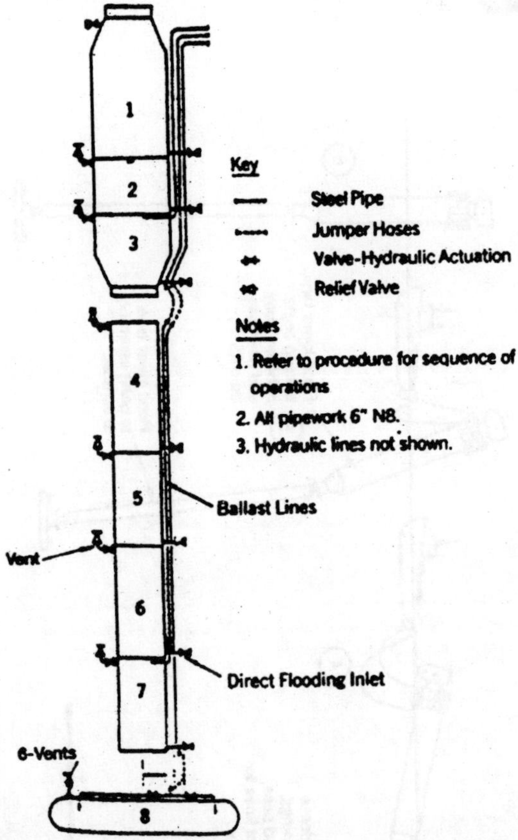

Figure 7.8: Installation of SALM: Bilge and Ballast System

platform to the base structure and a boat with bow thruster used to stretch them as the base reached the seafloor.

Where a floating production vessel (*e.g.* a converted tanker) is to be moored, a rigid but hinged yoke is usually fitted to the bow of the ship in a shipyard. The mating of the yoke and the single-point mooring riser is then carried out at the site, by use of a pull-in wire system, and connection made by a tension-bolted flange.

Another scheme involves vertical construction of the articulated column. The base is constructed first and moored in a deep-water protected site. The halves of the cardan are preassembled, joined

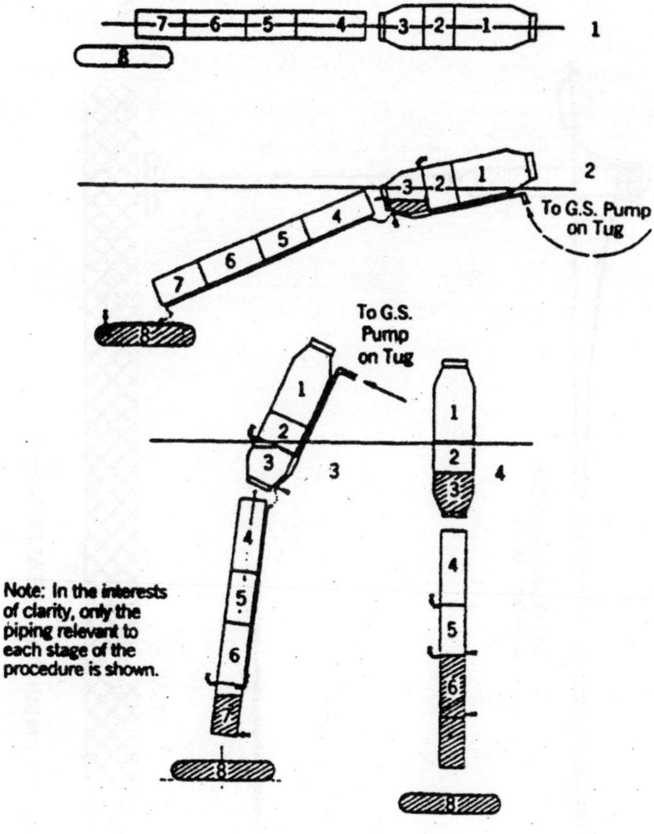

Figure 7.9: Installation of SALM: Uprighting Sequence and Flooding Diagram

with temporary fixing, and then set on the base. Temporary buoyancy tanks are attached to enable the base to be submerged and still remain stable. Then the column is constructed above the base, allowing the whole to sink slowly as weight is added. Eventually the structure will have been submerged to the point where the stability and the buoyancy can be supplied by the column alone. The temporary buoyancy tanks are then removed and the structure completed afloat. The deck can then be mounted, lifting it on, if it is a small structure, or transferring it by floating in over the column and deballasting, if it is a large deck. The whole can then be towed to the site in the

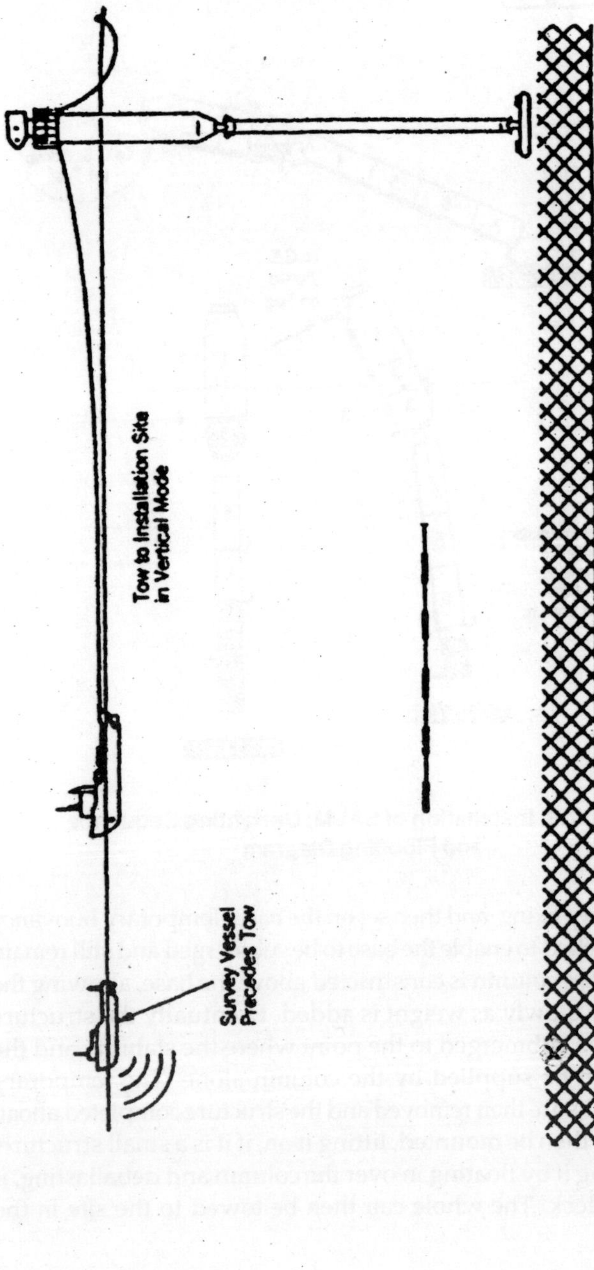

Figure 7.10: Installation of SALM: Step 7

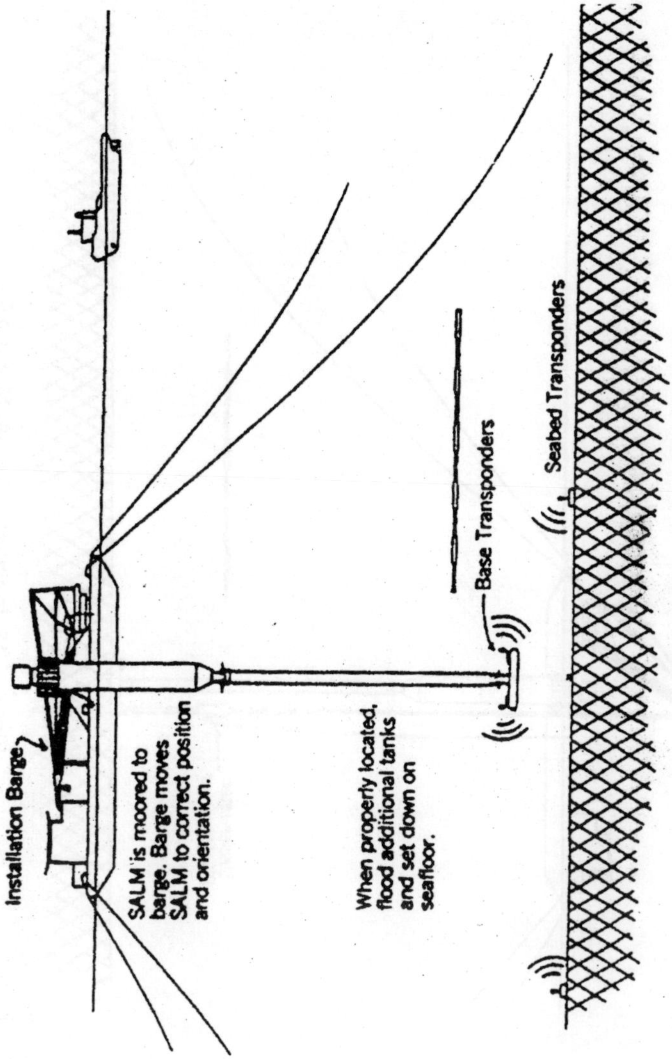

Figure 7.11: Installation of SALM: Step 8

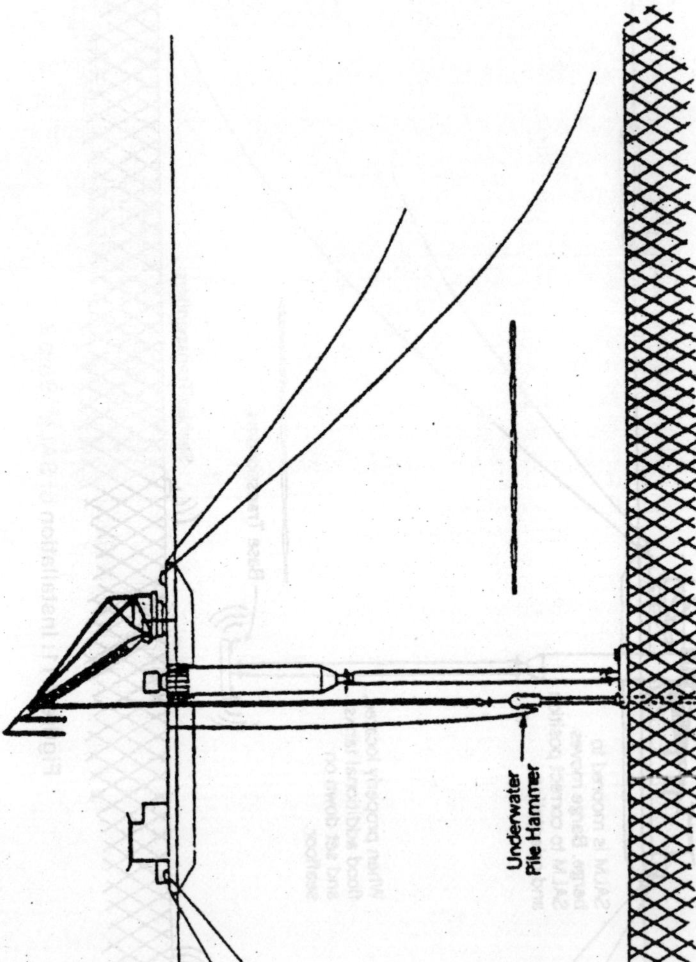

Figure 7.12: Installation of SALM: Step 9

vertical mode. After seating in the seafloor, the cardan is freed, to permit flexible articulation. Ballasting is added to cause the base to penetrate. In firm soils, underbase grout may be adequate; in other soils, underwater pilling will be driven through sleeves in the base and connected by grouting of the annulus.

The Use of Subsea Production Templates

The use of subsea production templates, working as part of a subsea production system, is growing rapidly. Even with fixed platforms (*e.g.* the BP platform Magnus) the use of a template through which to predrill the wells while the platform itself is being fabricated shows significant cash-flow advantages by enabling the platform to be brought into production at the earlier date.

Subsea templates have therefore been developed by a number of firms; they have grown in size and weight to as much as 2000 tons or more. These are typically loaded out on a barge for transport and then launched or lifted off to the self-floating mode.

They then must be lowered to the seafloor, with equipment and rigging adequate to sustain the high dynamic forces involved. Stability during the submergence is maintained by designing the structure so that the center of gravity is below the center of buoyancy.

In one system, the lowering is done by a drilling vessel, either a drillship or semisubmersible. Lines are run from the derrick hoist, down through the moon pool, and then up and attached to the top of the template floating alongside.

A floating derrick barge now hooks onto the template, which is given negative ballast, allowing it to sink below the keel of the drilling vessel. By slack the floating derrick barge's lines and taking in on those of the drilling rig, the load is transferred to the drilling rig. The barge lines are now disconnected.

This transfer can also be done dynamically without the use of the offshore derrick barge, provided the template has inherent stability, that is, center of gravity (CG) below center of buoyancy (CB). Buoyancy tanks attached to the upper portion of the template can be used to give this stability. For buoyancy tanks which must remain intact during descent into deep water, consideration should be given to offsetting the hydrostatic head by filling with a light fluid such as gasoline or solvent or by filling with syntactic foam.

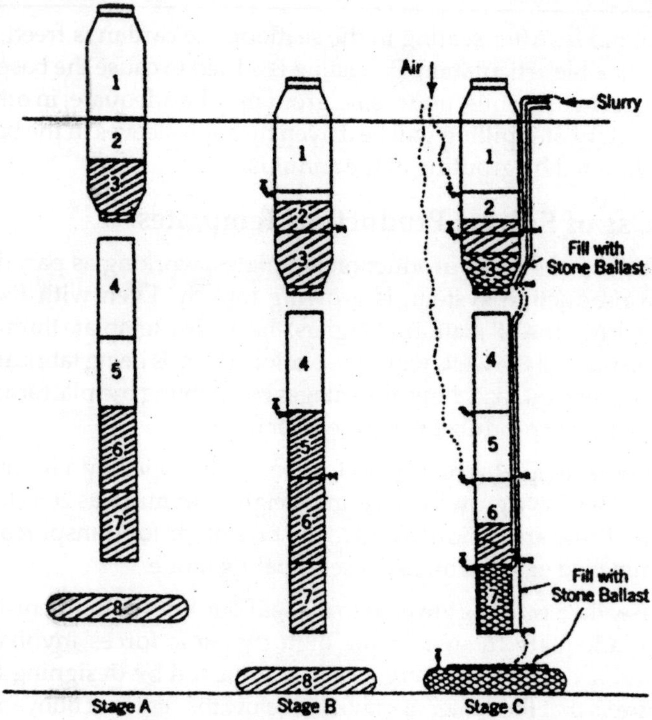

Figure 7.13: Installation of SALM: Setting Sequence

Other means are available for transporting and setting templates, especially the large ones. A barge may be specially equipped with large cable and casing grip hoists. In an inshore harbor, in relatively shallow water, the template is seated on the seafloor. The barge now floats in ever the top, attaches the cable or castings, pulls the template up snag under itself, and thus transports it to the site. Once at the site and properly positioned, the template is lowered to the seafloor.

Disconnect is effected by remote-operated devices, operated acoustically or hydraulically or by explosively activated disconnects. One advantage to the use of the drilling vessel to lower the template is that the heave compensator of the drilling vessel can minimize the dynamic loadings due to heave.

For the larger templates of the future, large spars can be used to provide inherent heave compensation as the template is lowered. In

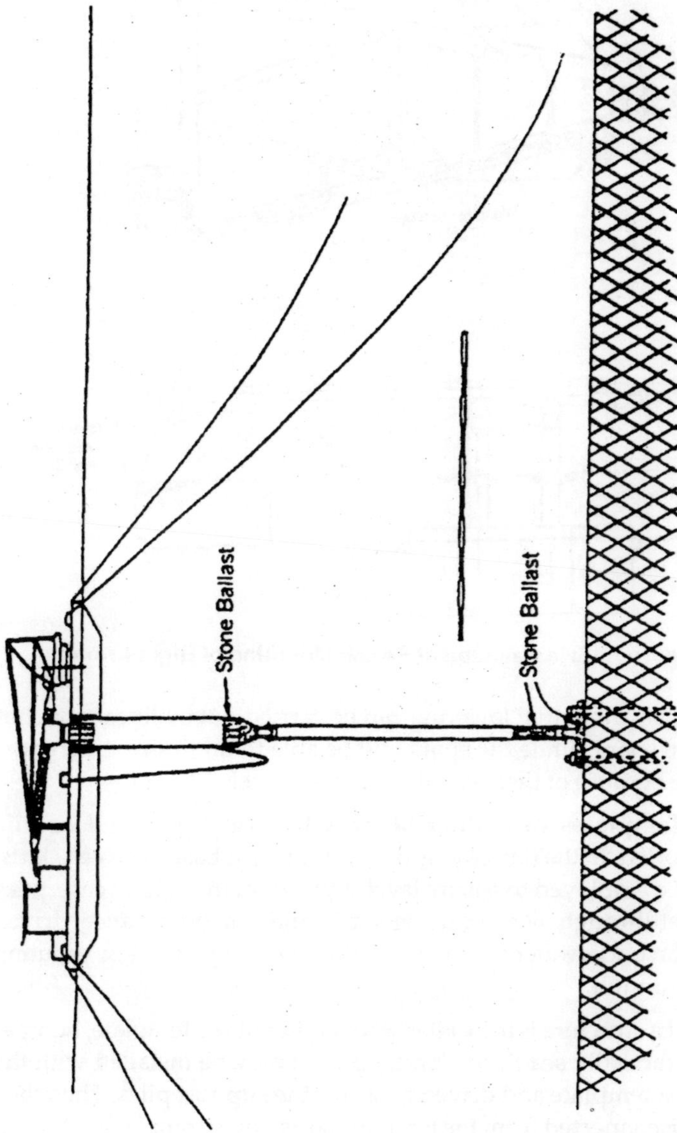

Figure 7.14: Installation of SALM: Step 10

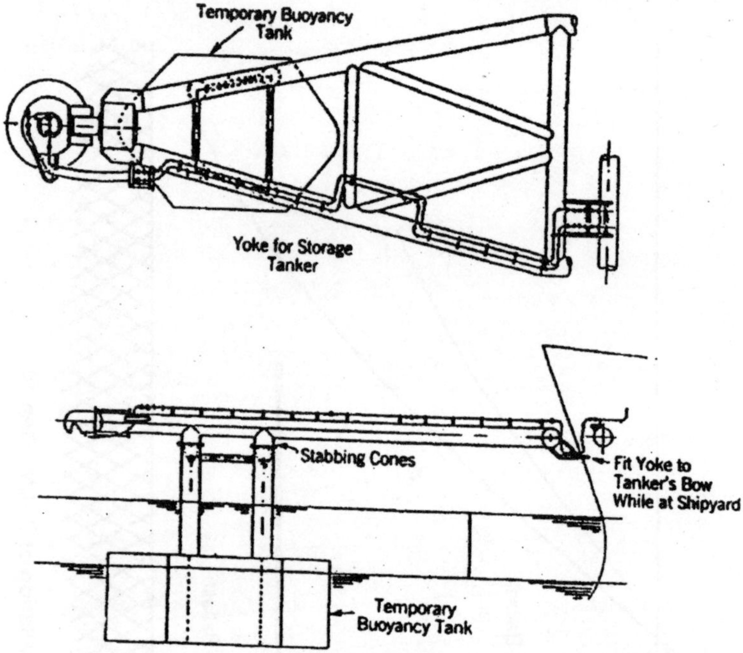

Figure 7.15: Installation of SALM: Mounting of Rigid Arm Yoke

this case, the actual lowering can be carried out by linear winches on the spars, while the spars will be able to minimize differential heave because of there small water plane area.

Templates can either be gravity base supported or pile supported. In the first case, grouting under the base, between skirts, may be employed to ensure level support. In the second case, piles are set through sleeves, using sonar and video guidance, driven with an underwater hammer, and the connection made by grouting of the annuli.

If a structure is to be later set over the subsea template, bumper piles (actually sea floor "fender piles") may be installed with the subsea template and driven in, as for the support piles. Then they are disconnected from the template so as not to transfer load into it during the structure's installation.

Flow line installation to subsea templates may be carried out by laying the flow line down beside the connection point. The line may

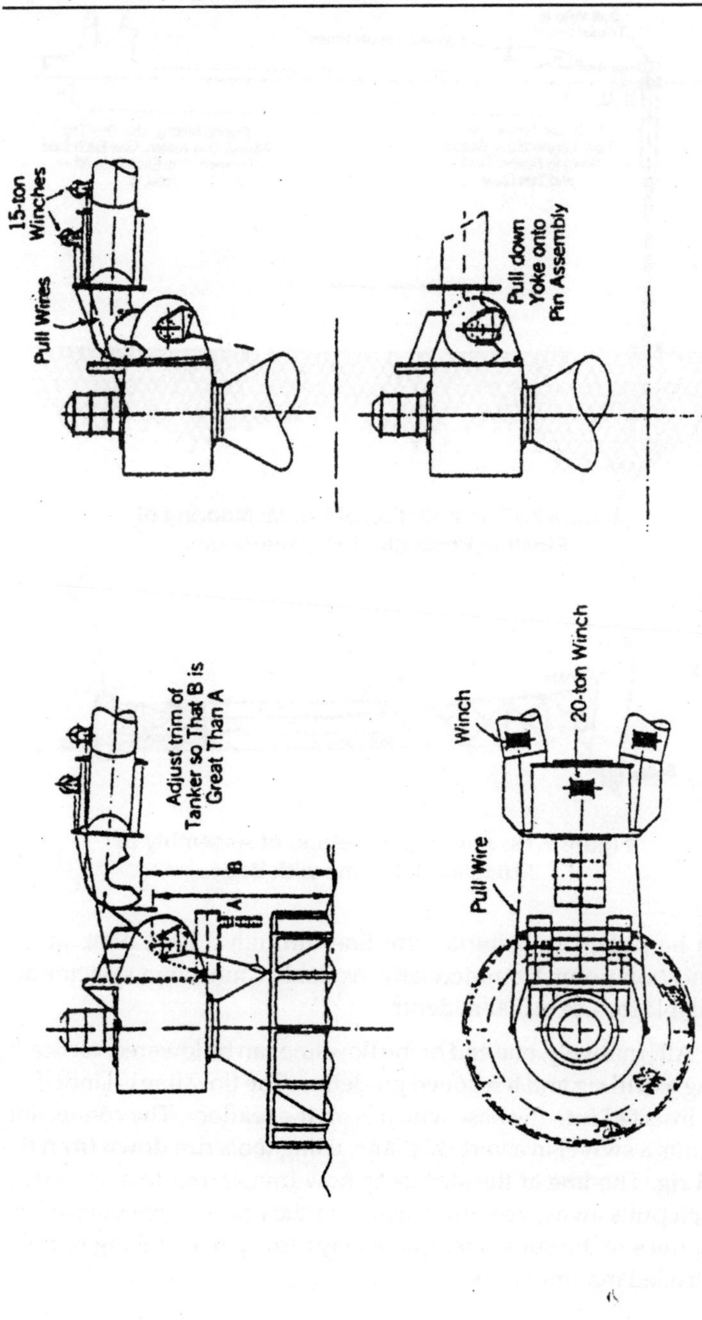

Figure 7.16: Installation of SALM: Details of Yoke Connections

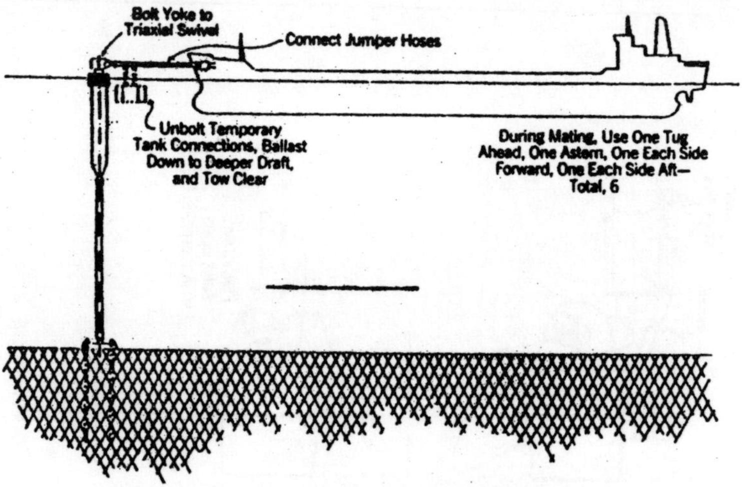

Figure 7.17: Installation of SALM: Mooring of Floating Production-storage Vessel

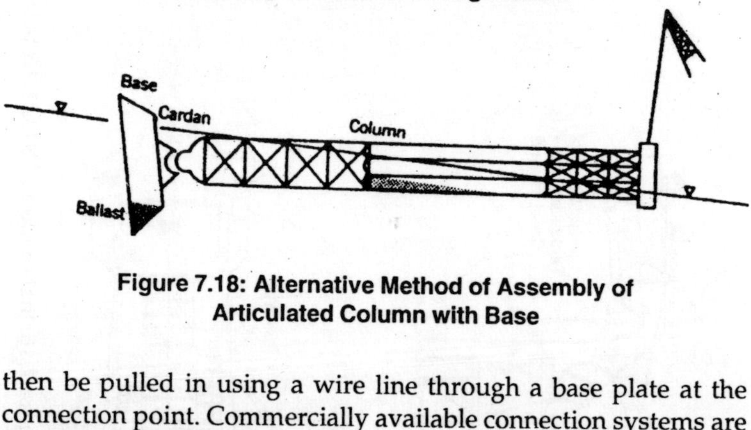

Figure 7.18: Alternative Method of Assembly of Articulated Column with Base

then be pulled in using a wire line through a base plate at the connection point. Commercially available connection systems are available for use to 300-m depth.

Alternatively, one end of the flow line can be lowered vertically using a drill rig and tensioned guideline. The flow line is landed on and inserted into the base which is on the seafloor. The connection permits a swivel in a vertical plane, using tools run down from the drill rig. The line at the surface is now transferred to a lay barge, which pulls away, keeping tension on the line. Connections of the flow lines to the subsea template may also be made using remote-controlled manipulators.

The Anchored Buoys in the Deep Ocean

It is often necessary to deploy anchored buoys in the deep ocean. These generally consist of a clump anchor, a large buoy, and a connecting line of either steel wire or more usually of fiber such as nylon, polypropylene, or Kevlar. To protect against damage to the line from fish bite, polyurethane or polyethylene coating may be extruded onto the fiber lines.

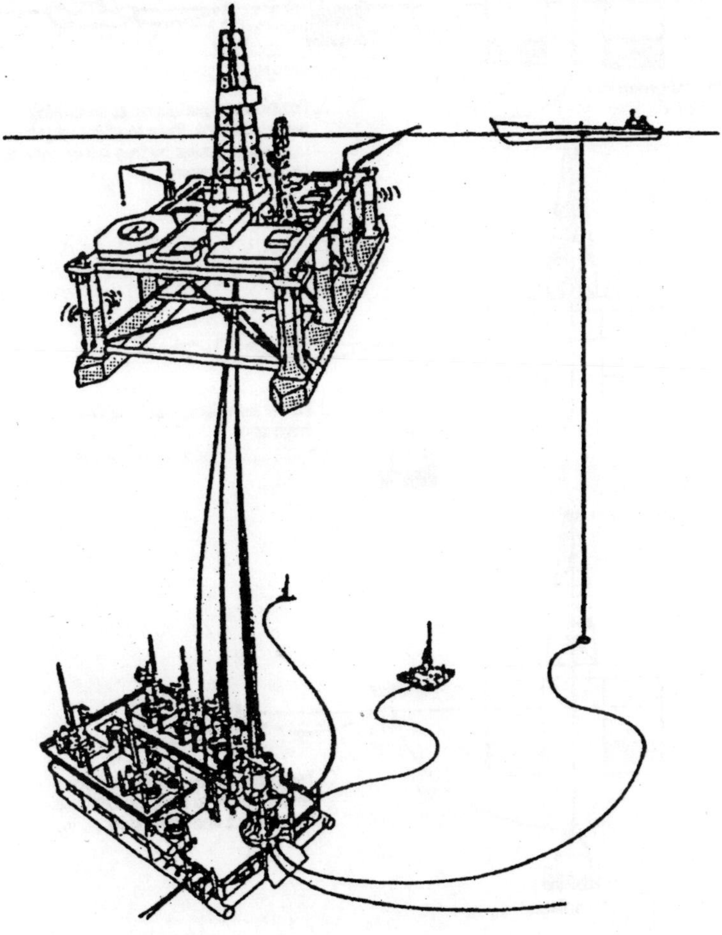

Figure 7.19: Multiwell Template and Semisubmersible Field Development

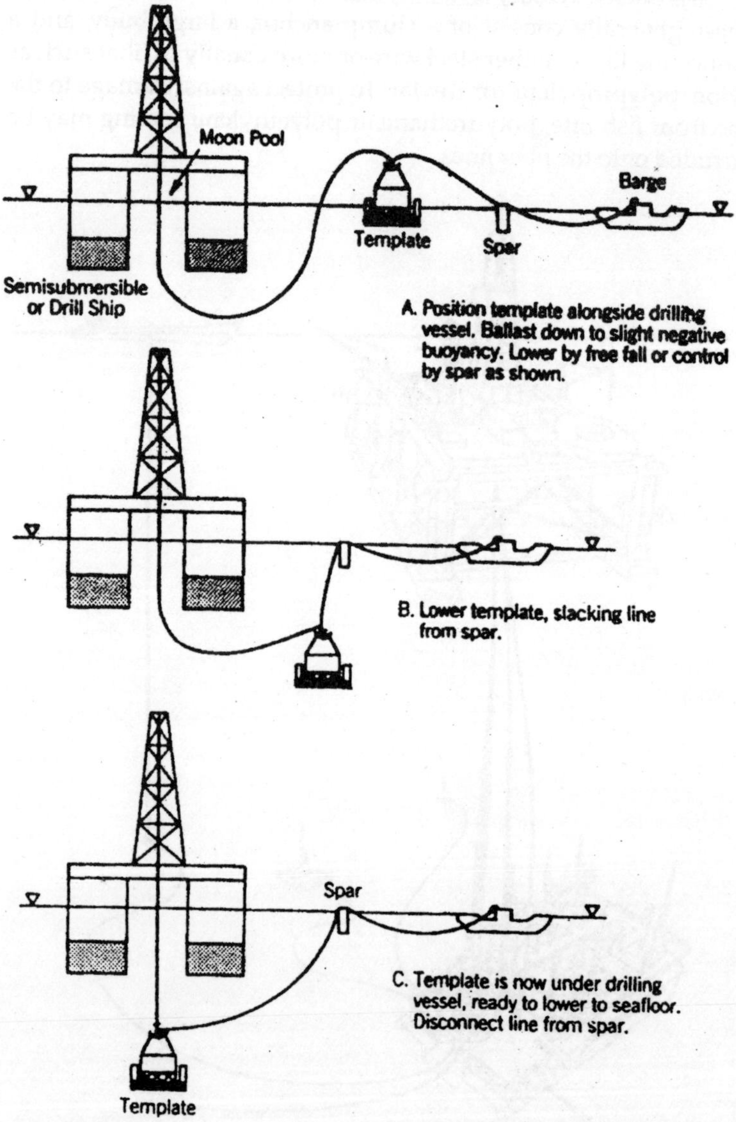

Figure 7.20: "Keel-Hauling" Method of Transferring Subsea Template from Afloat on Surface to Underneath Drilling Vessel

To prevent accidental sinking of the buoy from flooding, polyurethane foam or syntactic foam should be used as fill. Buoys have frequently been lost due to rupture of tanks due to boat collision, fatigue of welds, or gunshots from passing fishermen.

Generally the scope of the buoy line is set at 1.2 to 2.0 times the depth. Two methods of deployment are used: "anchor-first" and "anchor-last." In the anchor-first case, the anchor is hosted over the side and lowered to the seafloor, and then the buoy is deployed. This requires considerable time, during which position must be maintained as well as requiring a powered drum to control the lowering. In the anchor-last method, the buoy is launched, the boat travels forward a distance equal to the scope, and the anchor is dropped, free-fall. Both calculations and field measurements indicate that dynamic forces are relatively minimal, that the anchor falls only at 3 to 5 m/s due to the drag restraint of the line, and that impact forces of the anchor do not lead to excessive dynamic stresses in the line. In general, the maximum force in the line is equal to the weight of the anchor plus line.

When deploying cable arrays, relative position control is very important. In some cases, use of marker buoys (articulated spar buoys) with short scope may be helpful, although today with acoustic positioning systems and DGPS even more exact control can be attained. In some cases it is necessary to cut a line free–for example, a line used solely for deployment–in which case a hydraulic cutter can be run down the line, either controlled or free-fall and activated when it reaches the desired depth. Alternatively, an ROV equipped with a cable-cutting device may be employed.

Proving of the mooring capacity of a line may be carried out in relatively shallow water by mooring a barge, such as an offshore derrick barge, between the legs of the moor, then reeving wire line to blocks and thence to the deck engines, so that two legs reacts against each other. In deeper water, and with short scopes, the vertical component of force becomes dominant, and can be conservatively estimated to be equal to the underwater weight of the anchor or clump. Vertical capacity can also be proved by direct uplift.

Deployment of objects to the seafloor may be done by lowering down, as described in the section on seafloor templates, with adequate consideration for the dynamic response of the lowering vessel. Alternatively, controlled free-fall deployment may be used,

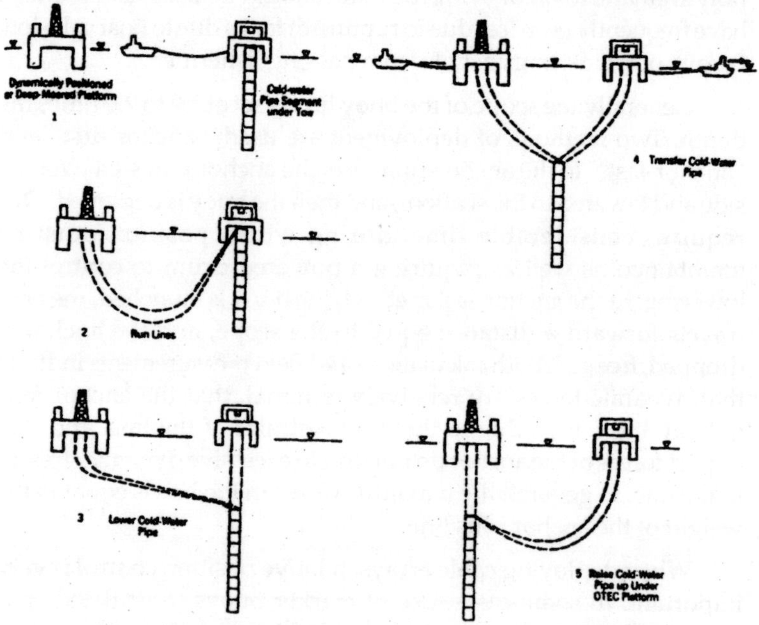

Figure 7.21: One Concept for Deployment of OTEC Cold-water pipe

in which buoys are used to reduce the net weight, increase the drag, and provide vertical stability. The buoys will thus also be progressively descending to deeper water, so that they must be filled with either light-density fluid or syntactic foam, so that they do not suffer significant decrease in volume with depth. Once the object impacts the seafloor, the buoy or sphere is released, either automatically on release of load (pelican-hook arrangement) or by an acoustic release mechanism.

Now let us try to understand about ocean thermal energy conversion.

Extensive studies and initial test installations for ocean thermal energy conversion (OTEC) have been carried out under contracts with the U.S. Department of Energy as well as by engineering organizations in other countries, notably Japan and The Netherlands. In general, all systems employ a cold water pipe of large diameter, raising cold water from a depth of 1000 to 2000 m to a surface plant where the warm waters are used to provide the heat source. The surface plant may be very large floating structure, in which case it is usually planned to be moored in deep water (2000 to 4000 m). Alternatively, the plant may be constructed on the continental shelf in perhaps 200 m water depth, or even on shore.

The principal unique construction aspects to be discussed in this section have to do with the development of the cold-water pipe. In test installations of relatively small-diameter pipe, several have been lost during development, while two have been successfully installed.

For the floating plant system, the pipe is usually conceived as being about 30 m in diameter, 1000 m in length, weighted at the low end, and suspended from the floating vessel at the upper end. Pipe materials considered have included light weight concrete, steel, fiberglass, polyethylene, and hybrid designs. The principal problems both in service and during construction are dynamic ones, resembling those successfully surmounted in the installations of the Cognac platform and Hutton and Heidrun tension leg platforms. Here, however, the sizes and weights are an order of magnitude greater. The major construction problems is that of transport, up-ending, and transfer of the cold-water pipe to its location hanging beneath the floating vessel.

The pipe is usually conceived as being articulated in order to reduce its in-service moments generated by internal waves and currents. During installation, the joints may be of use to the constructor or may require temporary locking. Alternatively, polyethylene pipes have been tried, which are able to be buckled during deployment yet subsequently regain there shape.

In one proposed system for deployment, long segments would be assembled in shallow water and towed to the site in the horizontal mode. Upon arrival, they would be upended by selective blasting. Temporary bulkheads or compartments within the pipe are required, along with filling and venting lines. Alternatively, the pipe may be placed under axial tension during upending, by means of tugs pulling in opposite directions.

Another deployment concept is that of on-site assembly in the vertical mode. As each segment of 50 to 100 m length arrived at the site, it would be swung under the vessel to the vertical position, winched in under the moon pool, then raised by the derrick so that the jointing with the previous segment could be made in the dry.

The successful use of high tension for upending the Heidrun TLP tethers is perhaps the most-promising solution for the OTEC cold-water pipe.

In recent years, most attention has been devoted to shelf-mounted plants, located on steeply sloping margins such as those found in Hawaii and on the east coast of Taiwan. For the MINI-OTEC plant on the island of Hawaii, a small polyethylene line ran down the slope to a depth of 600 m. Since it was positively buoyant, it was held down at intervals by tethers to clump weights. The line was assembled in quiet water near shore, towed afloat to the site, and submerged by progressively flooding. As it submerged, it progressively buckled but then resumed its circular shape after it passed the critical bend. Reinforcement and softening pads were provided at each tether to prevent chafing. The pipe was successfully installed.

Deep Oil Technology has proposed a system for pulling a large diameter line down a slope:

1. Using a drilling vessel, a pile anchor will be drilled and grouted into the rock at the lower end.
2. When the pile is set, it has a terminal gimbal base attached which contains a large sheave. Now, after the pile is in place, and cemented, a line is paid out from the drill ship, around the underwater sheave, and back to a surface boat. This end is towed to shore.
3. A landing base for the cold-water pipe termination is now lowered onto the pipe, using tensioned guidelines.
4. When the line reaches shore, it is attached to the prefabricated pipe. The pipeline is now pulled out to the termination pile. A small vessel is moored vertically above the end of the pipeline, moving out with it. The vessel supports a jetting hose and TV camera and provides buoyancy to support a drag scraper or plow.
5. Finally, a retrievable filter and entry port are lowered down to tensioned guidelines and fitted to the termination structure.

The Floating Structures and the Method of Launching

These are structures which are intended to remain floating throughout their service life. They include floating production, storage, and offloading (FPSO) vessels, the hulls of tension leg platforms and semisubmersible platforms, floating terminals, spars, floating bridges, industrial plants, floating dry docks, guide walls, and even the proposed floating heliports and airports. The proposed OTEC plant and the huge Mobile Offshore Base (MOBS) are also included. In many cases they involves the joining afloat of very large and massive segments. In service, they must endure the cyclic dynamic loading from waves and swells, so fatigue endurance is a significant consideration. Notable examples include the Heidrun TLP and the Troll West semisubmersible in the North Sea, the Sakti Ardjuna LPG-FPSO in the Java Sea, floating bridges in Norway and Washington, and the Valdez, Alaska floating wharf.

Some of the advantages inherent in concrete vessels and floating structures are long-term durability, with minimum maintenance, favorable thermal insulation properties and consequent lack of condensation, and due to their inherent massiveness, less response to waves, better resistance to abrasion and impact, and more freedom in location of superstructure and deck loads. They can be constructed in almost any country, using a preponderance of local materials and labor.

To attain these desirable qualities and advantages in concrete structures, prestressing is essential, as is well-distributed reinforcement. Concrete quality must be as high as possible in order to reduce draft and to obtain the impermeability essential to long-term durability. Because of their intended long-term service in a dynamic environment, the levels of prestressing and the quantity of passive reinforcement must be adequate to ensure that after crocking of concrete, the steel is below yield. The sections are typically thin as compared with offshore concrete platforms. Therefore, tolerances become of increased importance, both for the concrete sections and for the placement of reinforcing steel and prestressing ducts.

Many large floating structures, *e.g.* FPSOs, spars and the hulls of TLPs, are made of steel. These structures are usually fabricated in a construction basin or graving dock but some have been fabricated on ways. Smaller vessels may be fabricated on a large barge. In all cases, draft at launching may well be critical, so very accurate control of as-built weight and external dimensions must be maintained. Weight is affected by the concrete density, and the weight of reinforcing steel actually placed, including splices.

Hull forms are typically rectangular with a flared bow and tapered stern, since the towing and service wave responses are primarily inertial. Being deep in draft, due to their mass, they typically respond less to waves and swells than their steel counterparts. For TLPs and semisubmersible hulls, spars, OTEC and other deep draft structures, the cylindrical or oval cross sections are much more efficient, due to the ability to use the entire cross section in compression to resist hydrostatic loads. Concrete, being readily molded to any desired shape, is also excellent for such complex shapes as the elbow of TLPs and the torus.

Both standard-weight concrete and light-weight concrete are employed. The lightweight concrete, with a saving of 20 per cent in weight and hence draft, may require 15 to 20 per cent additional reinforcement in both in-plane and out-of-plane shear. However, where the plate thickens is determined by cover and the need to properly space the inner and outer reinforcements, as well to accommodate ducts for prestressing, the lighter unit weight may result in more efficient and practicable vessel. Lightweight concrete further possesses better thermal properties and, when combined with concrete containing silica fume, increases fatigue endurance. It behaves especially well at cryogenic temperatures.

Steel floating storage barges and FPSO vessels are extensively deployed worldwide. They are built by conventional shipbuilding methods. Many are turret-moored.

Maximum utilization should be made of prefabrication. Typically, joints are cast-in-place to ensure full continuity of reinforcing steel and to permit splicing of ducts. Precasting permits the attainment of dimensional accuracy, while dispersing construction activities and speeding schedule.

The base slab is laid or cast on a carefully screeded crushed rock underbase, to permit later flooding to penetrate entirely under the rock. This also permits attachment of any fittings which must protrude below the hull proper. In the case of a cast-in-place bottom slab, the rock is covered with polyethylene or similar. Drainpipe is placed in the rock underbase.

Base slabs has also been successfully cast on a finished concrete slab which has been coated with bond breaker. The base slab is typically haunched upward, 100 to 150 mm or so at the wall panel intersections, so that the eccentricity of the prestressing tendons will prevent the negative moment tension cracking.

Wall segments are fabricated slabs, tilted up using tilt-up building technology, and joined, usually at the juncture of four wall slabs. While this is a very congested area, it may be enlarged by 45° fillets. The side walls should be haunched over the frames.

What appears to be the optimum system for joining the wall slab to the base is to leave a space in the base slabs, and design the joint in the base slabs at the same location as the joint with the wall. The wall slab now is constructed with its reinforcing steel projecting into the joint, as are the two slabs. Thus the bar needed to be set in exact location so that the mating bars can pass without interference. The wall slab is supported by struts at its sides which rest on the precast slabs. Prestressing ducts are spliced and then the concrete joint is placed, integrating the haunches previously described.

Mechanically headed rebars are especially useful in regions of high shear, since they can be placed more readily and effectively in the congested areas. Mechanical rebar splices are not only better from the point of view of congestion but reduce weight.

The upper deck may similarly be constructed of precast slabs, resting on ledges bolted to the walls. The joints then are made over the walls.

Precasting is especially useful with double-curved portions of the hull, such as those for the bow. Slip forms have been used for cast-in-place walls. Stay-in-place forms have been used for horizontal members such as the deck.

Construction joints should be properly prepared by water jet blasting to expose the coarse aggregate. Typically, this will cut back the joint surface about 6 mm. Then just before the concrete is placed, bonding epoxy may be sprayed on the joint surface. It must still be wet as the concrete covers it. Some experienced fabricators prefer to brush on the heavy coat of mortar or latex mortar.

Decks should be sloped to drain so that saltwater spray does not pond. Decks should be treated with silane at the time of construction and at 2 to 5 year intervals thereafter. In vessels scheduled for Arctic and sub-Arctic services, air entrainment should be used in the concrete to prevent freeze-thaw attack.

Anchorage zones needed to be well reinforced to distribute and confine the transverse, longitudinal, and radial strain which occur. Although this is matter of design, these details are often made the responsibility of the contractor or the prestressing subcontractor, since they depend somewhat on the anchorage details selected. Further, since many of these large floating structures are constructed on a design-build basis, the constructor is vitally involved. The highest stresses around the anchorage typically occur at the time of prestressing, not under service or extreme loads, and it is then when cracking occurs if proper details are not employed.

Post-tensioning ducts should be of plastic, fused at their splice joints. Grout should contain an antiwashout and thixotrophic admixture to prevent the formation of bleed pockets. Plastic caps should be used to seal over the anchorage. Prestressing anchorages should be protected by being located in a pocket, into which reinforcing bars protrude from the surrounding concrete structure. This is filled with concrete after all prestressing activities are finished. The joint surfaces should be coated with bonding epoxy just before placement of the concrete. In the Arctic and sub-Arctic, latex concrete should be used instead and the epoxy coating eliminated.

Mooring fittings such as posts and bollards should be post-tensioned to a thickened and heavily reinforced seat in the hull. Usually prestressing bars will be used instead of strands, since seating losses of prestress can be eliminated by the use of bars.

Fairleads and/or hawse pipes are either post-tensioned on or embedded in the concrete with proper anchorage. Steel plating should be placed to eliminate the wear from the rubbing of anchor chains. Piping and mechanical systems should, as far as is practicable, be integrated into the hull design, with embedments pre-installed. Corrosion protection should be provided with sacrificial anodes, especially at sea-water intakes and discharges.

There are many other design and construction details applicable to concrete vessels. Reference is made to Gerwick *Construction of Prestressed Concrete*, 2nd edition, McGraw-Hill, New York, 1996.

An especially appropriate hull design has been developed as a result of testing at both the University of Manchester, U.K., and the

University of California at Berkeley. It is a double-steel shell, made to work compositely with concrete infill. Thus it can be designed for shell action to resist high hydrostatic pressures and, at the same time, benefit from the ductile behavior of the steel. Additional testing of the concept for ice-resisting structures has been carried on in Japan.

Constructibility of these large vessels requires consideration of access and crane reach, since work will need to be pursued concurrently at many locations.

Now let us learn about method of launching of floating structures.

The most usual method of launching is by flooding of the basin or dry dock so that the structure becomes waterborne. Exit is controlled both by lines to the dock and by tugs. To break the hull free from a casting slab or forms, water is introduced underneath the base slab. Low pressure, sustained for several hours, is best. It is most effective if injected through plastic pipes embedded in the hull, which are later grouted.

Launching on a ways is done by side-launching, in order to reduce bending moments and concentrated loads on the way. A concrete barge for a floating phosphate plant was built in Singapore. It was constructed at grade, on the edge of harbor, then jacked up and slid down the ways into the water. As the barge leaves the ways, it raises off on one side but the other side may put a temporary concentrated load on the ways. Similarly, the change in support causes high transverse bending in the barge.

Launching from a barge hull is the most demanding. Stability must be assured during both the sinking and the offloading. Stability of the launch barge must be assured after the new barge floats off and while the launch barge floats back up. This is often done by provision of columns which extend through the water plane. When the barge raises off, if it does so at an angle, it may load the vessel's deck with a concentrated load. Safety needed to be verified.

Semisubmersibles are much more difficult to build and launch. If the semisubmersible has large pontoon hulls, then each of these two hulls can be constructed by one of the methods used for conventional barge hulls. Then after they launched, the shafts are constructed. The two hulls are now brought into proper position relative to each other and connected by bracing, both at the level of the hulls and at the top of the shafts, as well as with crossing diagonal ties. The double-hull structure may now be ballasted down to facilitate the construction of the upper-deck structure. The bottom hulls may even be seated temporarily on the seafloor if it has been properly leveled.

Even more difficult is the construction of semisubmersible structures which do not have substantial bottom hulls. This type of configuration has been proposed for some floating airport structures and some designs for MOBS. In one erection scenario, individual lower modules are properly positioned adjacent to each other in a construction basin or graving dock. Then the basin or dock is flooded. A previously fabricated upper deck is now floated over the shafts, temporarily supported on a barge or barges. Then the barges are ballasted down so that the connection can be made between shafts and upper deck, in much the same way that the deck of a gravity-base offshore platform is connected to the shafts. Post-tensioning is a rapid means of proper connection and may be supplemented by welding in the case of steel connections.

Prestressed concrete and especially prestressed lightweight concrete have favorable properties in that the concrete and prestressing steels are not subject to embrittlement at low temperatures but just get slightly stronger. The thermal conductivity through the thickness is low, and fire resistance is high.

However, the fact that the concrete has through-thickness insulating properties means that it will have large thermal differentials and hence strains. The cold face will develop tension and the warm face, compression. At corners, especially where slabs join in three orthogonal directions, severe strains will arise, typically results in cracking. Special reinforcing and/or prestressing details are required. Plastic insulation is generally provided, along with a thin stainless steel membrane to prevent the penetration of liquid gas into the concrete. Insulation on the inside eliminates and/or minimizes these thermal problems. Tanks wall exposed to the cold will, of course, shrink globally, producing through-thickness strains. If these tank walls are also subjected to axial tension and compression due to waves, they will need to be prestressed to prevent fatigue, since the cracking lowers the otherwise excellent fatigue endurance.

The permanently floating steel structures have been used for many years as floating oil storage vessels. Design and construction follows shipbuilding practice, except that, being stationary vessels, they are not faced with necessity to reduce drag; hence they can be very simple large boxes, with vertical sides and ends.

FPSOs of steel are permanently moved in the North Sea and the Middle East. Designed to weathervane to minimize wave forces, they are molded like typical seagoing vessels. Most recently, turret moorings have been employed to weathervane about their center of rotation and top permit oil and product lines to be swiveled in the turret.

Recently, the spar has re-emerged as a potentially optimum vessel for offshore oil drilling, production, storage, and offloading in the very deep ocean, subject to severe storms. These are much larger than previous spars and have drafts of 150 m or so. They are built in shipyards in the horizontal position and launched like a ship. Concrete versions of the spar have been proposed, utilizing the favorable ability of concrete to resist hydrostatic loads.

Chapter 8

Offshore Structure Repairs

Introduction

Offshore structures are exposed not only to the extreme conditions of the environment such as wave slam, ice impact, and fatigue, but also to accidental events such as boat impact and objects dropped off the platform.

The list of such accidental events that have occurred over the past several decades is myriad. It includes ramming by a supply boat that went full ahead rather than full astern, impact from the reinforced corner of a cargo barge, and impact by a derrick barge whose mooring lines had parted. The dropped-objects category includes a number of pedestal cranes pulled off their supports when they attempted to follow the movements of a supply boat and thus exceeded their allowable radii, also drill collars, casing, a mud pump, and pile hammer. Anchors have been dragged across a pipeline. Leaks into underwater compartments have developed through corroded and ruptured piping.

In the environmental category, horizontal bracings near the waterline have been excited by vortex action and subjected to vertical cycling beyond that for which they were designed, with consequent failure by fatigue. Scour has undermined a pipeline, to develop long, unsupported spans, which then ruptured under the cyclic loading due to vortex shedding. Defective welds and heat-affected zones have led to crack development, and its subsequent propagation has been accelerated by corrosion in the crack. A platform may also be damaged by operational failure, which leads to flooding or overloads. Scour may undermine the legs and lead to excessive lateral response. Finally, corrosion may occur which weakens the structure beyond allowable limits.

Thus repairs become necessary. Obviously, each such undertaking is highly case-specific and needs to be engineered appropriately to the particular needs of the case.

One fundamental principal is that the carrying out of the repair must not increase the risk of failure. This may require auxiliary strengthening before the damaged member is cut out for repair. It may mean that repairs must be delayed until a more favorable season, when the environmental loads are reduced. It may mean that limits are placed on operations until the repairs are completed. Repairs may have to be carried out step by step. A second principle is that the repaired or constructed element must not adversely affect the performance of the structure. As an example, if the reconstructed member will be significantly stiffer than the original member, a dynamic reanalysis of the entire platform may be required.

In order to reduce the time of repair to a minimum, not only to limit the cost but especially to limit the time when the platform is in a weakened condition, the repair procedures must be planned in extreme detail and all necessary tools, rigging, and fittings provided as one package. For complex repairs, a rehearsal with the crew may be advisable to reduce the problems of communication during the actual repair.

Many repairs are located near the sea-air interface, and hence the repair work will have to be carried out under conditions of wave turbulence due to incident, refracted, and reflected waves. Some local protection may be able to be provided by the derrick barge acting as a floating breakwater.

A major problem with large offshore structures is that of identifying locations so that divers, ROVs, and submersibles may readily return to specific spots. The pre-installation of large, highly visible (yellow or orange) numerals on platform legs and walls is now routine practice. However, for a specific repair, additional local markers must be installed. Attachment of a wire guide line as a first step will facilitate descent and location and help the diver to maintain position even in a strong current.

Cleaning of marine growth is another operation which must be carried out at an early stage. High-pressure water jets are the most effective means, although in order to thoroughly inspect or commence repair of the crack in a weld, for example, supplemental wire brushing may be necessary.

Inspection can be carried out by divers, a diving bell, or an ROV with video. The latter two are replacing much of the diver work in order to reduce costs and increase safety. Especially when it is necessary to inspect inside the jacket frame, the ROV will eliminate the potential for divers to be trapped. When divers must go into a frame or under a structure, at least two divers work as a team, with one tending the lines for the other. As a first operation, a guide line can be carried in and attached, enabling the diver to easily retrace his path.

When corrosion is the cause of damage, the repairs must include a determination of the cause. Steps, such as installing sacrificial anodes to provide cathodic protection, must be instituted to prevent on-going corrosion.

Repairs to Damaged Braces

When diagonal or horizontal braces have been fractured or badly distorted, the damaged sections can be cut out. By use of a template and careful measurement, a "pulp" is now cut to exact length and end profile, with the ends beveled for full-penetration welding. External clamp are used to hold the pup in position. Now a habitat is placed around the brace and dewatered, and the full-penetration welds are made. *See* Figure 8.1. Alternatively, underwater "wet" welding may be employed, provided tests show that satisfactory quality can be attained.

Another solution to this problem is to cut out the damaged portion of the brace and then slip in a longer section of slightly smaller-diameter but thicker-walled pipe to which packers and grout fittings have been attached. Alternatively, the pipe may be of larger diameter and slipped over the two undamaged ends. This inserts pipe may be made in sections of threaded casing, to enable a longer length to be inserted or slipped over or through a small cutout. By

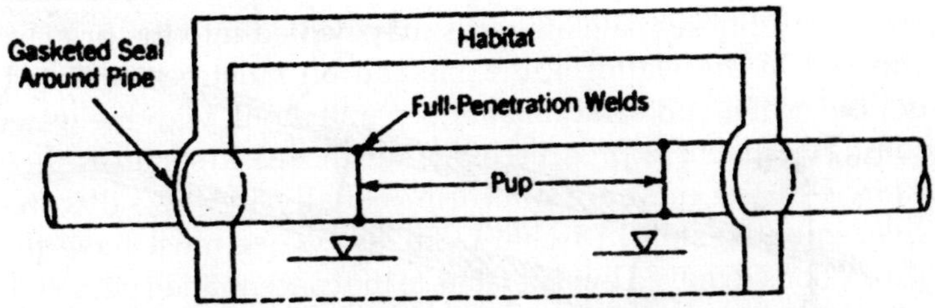

Figure 8.1: Repairs to Damaged Braces: Scheme 1

the use of cementing techniques similar to pile-to-jacket sleeve connections, the strength of the brace may be restored. The packers are inflated at each end and the grout injected and vented into a standpipe at the high end to ensure that all water has been ejected. *See* Figure 8.2. A third scheme uses both an internal and external sleeve. Only the external sleeve is structural. The internal sleeve is merely an internal form to facilitate grouting. *See* Figure 8.3. Special grouts are used, having expansive and high-bond capabilities in order to develop the transfer in as short a length of overlap as possible. The provisions of API-RP2A concerning grout transfer from pile to sleeve are followed as a guide. Multiple weld beads or shear keys are used to enable grout bond-shear transfer in a short length. Epoxy injection is an alternate method.

Another scheme involves the insert of a heavy internal structural member of smaller diameter than the original tubular. This new section can be a heavy-walled tubular with shear rings welded on the ends or even a rolled shape, for example, an H pile. At the lower

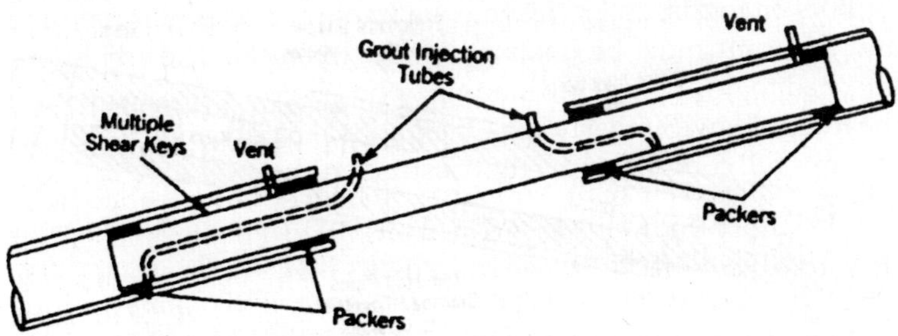

Figure 8.2: Repairs to Damaged Braces: Scheme 2

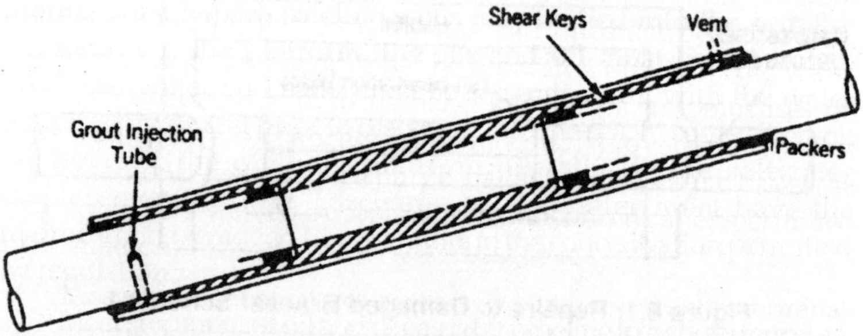

Figure 8.3: Repairs to Damaged Braces: Scheme 3

end, a pig or packer is attached. An external sleeve is clamped over the gap, using two half sections, with flanged and gasketed joints. Then the brace is pumped full of concrete. The concrete is actually a fine concrete, using cement plus sand, with the sand graded up to about 6-mm maximum size. *See* Figure 8.4.

Hydraulic expanders are in the process of development, primarily for use in expanding piles against jacket sleeves to transfer the load by a combination of direct shear on the corrugated surfaces and friction, thus supplementing or replacing load transfer by grout. This "swaging" process appears especially useful for repairs of damaged tubular bracing. It is, of course, to be noted that all these schemes are but variants on one scheme, with effort directed toward simplicity and reliability.

Other schemes have been used, especially where axial loads are light, involving clamped external sleeves. These may be in two

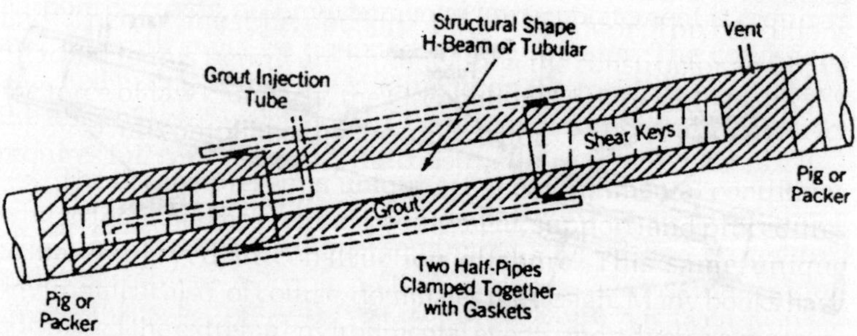

Figure 8.4: Repairs to Damaged Braces: Scheme 4

halves and use high-strength bolts to draw them in tight against the arms of the original brace. Load transfer is by the friction of steel on steel.

In the late 1980s Chevron reported the discovery of a serious fracture in one of the underwater tubular braces of the Ninian Southern platform. The fracture was discovered by an ROV during routine inspection. The fracture actually was a complete severance of the tubular and is believed to have been due to a faulty weld on a fabrication window used to gain access for a backup weld at a node. It is interesting to note how many serious fractures in tubular have occurred not at the critical nodes but rather at temporary closures or attachments: the Alexander Kjelland tragedy was due to a faulty weld in attaching a sonic transducer.

Repairs were carried out by Chevron at a depth of 43 m. the tubular was 1200 mm in diameter, fabricated if 30 mm plate. Because the brace would have to take cyclic tension, the decision was made to use a welded repair. It was feared that a bolted and grouted connection might not have the necessary fatigue resistance under the required stress ranges and number of cycles involved.

Oceaneering, Inc., built a specially designed underwater habitat in which a team of saturation divers would work. The fractured faces were removed and sent for a metallurgical examination. The two ends of the braces were now prepared for the weld. Then a lead-filled template was fitted to each end of the tubular to get an exact impression of the exposed ends. From these impressions, transition pieces of 500-mm plate were fabricated and then welded in place, leaving a gap between them to enable access for a backup weld on the insides. Special collars had been welded around the transition pieces. High-strength bolts were now tensioned to draw the collars together with 400 tons of tension force.

A 50-mm "pup" (a closure length or spool) was now fitted into the gap and welded from the outside. The cooling of the weld added to the pretension in the brace, so that the final stress was 600 tons of tension, thus restoring the state of stress that was originally in the member under static load conditions. *See* Figure 8.5. All welds were inspected and platform was recertified.

Cracks in the welds may be treated in a number of ways, depending on their size and extent, the service required, and the

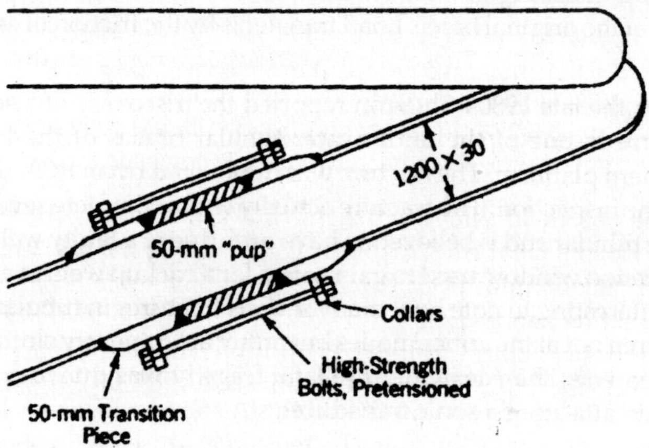

Figure 8.5: Schematic Representation of Underwater Repair to Tubular Brace of Ninian Central Platform

cause of the fracture. In any event, the external surface must be ground smooth. After inspection, if the crack is in the non-severe category, small holes may be drilled at each end to act as crack-propagation arrestors. The holes and crack may now be filled with epoxy, using crack-injection techniques. More serious cracks must be gouged out and re-welded, using either a habitat or wet-welding techniques. Cracked nodes have been repaired by clamping on the external nodal sleeves made up of as many sections as required and injecting the space with epoxy. In some minor cases of dented jacket bracings or legs, it may be sufficient to fill the legs with grout to inhibit compressive buckling.

As with steel structures, concrete structures may be damaged by impact, collision, and dropped objects. They may also be damaged by over- or under pressurization beyond that for which they were designed, resulting in flexural and shear cracks. Excessive foundation settlement may lead to shear cracks. Repeated cycling at high stress ranges, especially if they extended into tension, may cause low-cycle fatigue of the concrete or reinforcement. Marine structures in a seawater environment are subject to corrosion of the reinforcement, to freeze-thaw attacks, and to chemical disintegration.

The Beryl A platform was severely cracked during construction at the deep-water moorings, by ballasting which overlooked the temporary differential in heads on the numerous small interstitial cells, called star cells. This led to excessive tension plus shear in the cell walls, resulting in through-wall cracks and displacements of the star cell walls. Repairs were carried out expeditiously and adequately using the following techniques.

Removal of cracked and crushed concrete by chipping by hand or with a very small air hammer so as not to damage further the adjacent concrete;

Where major displacement of the walls had occurred (up to 100 mm radially), the addition of an inner reinforced concrete wall about 200 mm thick, tied to the existing wall with drilled-in and grouted studs and expansion bolts;

Use of grout-intruded aggregate as a means of correcting wide cracks and strengthening walls;

Use of cement grout injection in moderate-width cracks;

Epoxy injection of all cracks, carried out by selling the cracks externally and then injecting a fluid epoxy until material of a similar consistency flowed out the next port above. The cracks were successfully injected against an external water pressure that ranged up to a 40 m head;

Shotcrete application to build up the shell walls for compressive resistance.

Numerous cores plus acoustic testing verified the soundness of the repaired structure, and it was therefore certified for installation. The platform was installed successfully, despite having to sustain differential heads up to 100 m. the repairs have shown no problem in over 25 years of service.

When the Statfjord A platform was ballasted down to mount the deck, a small leak was noted in a juncture wall with another cell. Attempts to epoxy-inject this crack were less successful than with Beryl A. In the case of Beryl A, the cracks were through the wall and hence were crossed by reinforcing steel. In the case of Statjford A, the crack was laminar and had no reinforcing across it. Eventhough the injection pressure was limited, the act of injection propagated the crack. A decision was made to limit injections and to fill the adjoining star cells with a long-term corrosion-inhibiting gel.

It would appear from the experience at Statfjord A as well as that on subsequent long-span bridges, that stitch bolts should be drilled in and grouted before attempting to epoxy-injects a laminar (in-plane) crack.

One of the shafts of a North Sea concrete platform was impacted by an anchor-handling barge whose heavily reinforced corner caused significant cracking at and just below the waterline. Some spray leakage occurred. Repairs were successfully carried out by use of an external work caisson lowered on the exterior of the shaft, clamped to it, and dewatered. The damaged concrete was then cut out and additional reinforcing bars installed. New concrete was then placed as grout-intruder aggregate, using a colloidal mixing process to increase the fluidity of the grout. Epoxy was injected into the construction joint to fill any shrinkage cracks; very little "take" was noted.

The loss of under pressure in a concrete oil storage platform may lead to cracking. Essentially this is an overpressure situation in which the oil inside, being less dense than the water with which it is automatically equalized at the bottom, exerts an upward pressure on the roof of the storage caisson, causing multiple horizontal cracks in the cell walls just below their juncture with the roof. The Draugen platform experienced some leakage during storms due to the pumping action in the cracks.

In many cases such cracks should be able to be repaired by epoxy injection, applied by divers from the exterior periphery of the base caisson. However, in the case of Draugen, the cracks were active cracks, that is, only open during cycles of wave action, but closed during the passive sea states needed for repairs. Hence, only partial sealing was able to be effected and some cells had to be taken out of service. Access to star cells may be very difficult, and, depending on the circumstances, other means of crack stopping may be preferred. For example, depending on a detailed evaluation of the design, it may be feasible to fill star cells with a stabilized or gelled drilling fluid, which will prevent leakage and future corrosion.

Repair procedures for potential major damage to concrete structures have been developed and tested by Italian Engineers in connection with the approval of the Genoa floating concrete dry dock. These included the case of severed prestressing tendons, for which the repair procedure consisted of cutting out the damaged concrete, splicing the severed tendons, and restoring the prestressing force by internal jacks.

More recently, Taylor-Woodrow demonstrated a means of repairing severed tendons underwater. The damaged concrete was removed by high-pressure water jets operated by divers. The damaged and distorted reinforcing bars were cut out by an oxy-arc-burner. Broken prestressing tendons were then coupled to prestressing bars. A prefabricated reinforcing grid was then placed and tied to the existing steel. Prefabricated formwork was secured and gasketed to the existing structure. Then a high-strength concrete mix was pumped in from the bottom. After curing the concrete, the formwork was stripped and the bars stressed and grouted. Tests showed that the structural strength was fully restored.

External prestressing has been successfully used to repair damaged bridge girders and would appear practicable for use in underwater structural repairs as well. Holes drilled into structural elements could be used for the end anchors. Polyethylene-encased prestressing strands would have long-term durability.

Leakage has occurred in the base of the Beryl A platform, with water penetrating along nests of small-diameter grout piping. Cement grout injection in stages, using techniques developed in tunneling, has proved successful.

The recent development of carbon fiber "patch" and "strip" repair systems appears of particular importance for concrete marine structures. The patch is "glued on," using epoxy. Provided the fibers are properly oriented, the patch adds stiff reinforcement on the face. "Flexible" patches of suitable composite materials can be "glued on" underwater to restrain crack propagation and leakage.

The permeation of chlorides into concrete depassivates the reinforcing steel. The concurrent penetration of oxygen then leads to corrosion. Problems are especially prevalent in the "splash-zone" extending from low water to above-normal wave action, but can also occur well above high water due to airborne chlorides. The corrosion of the steel causes cracking and spalling of the concrete.

Standard repairs consist of removing the concrete from around the affected reinforcing bars, cleaning the bars, coating them with zinc silicates or epoxy, and placing new concrete. While this method gives short-term improvement, corrosion often re-emerges within a short time in the peripheral areas surrounding the patch. Provision of the zinc strip anodes, attached to the end of each bar, appears effective in preventing the spread of the corrosion. Corrosion below water is rare, due to the low oxygen concentration in seawater and the limitation of the chloride ions to those in sea water. If corrosion does occur, sacrificial anode cathodic protection is effective, practicable, and economical.

Impressed current cathodic protection is effective but only if it is properly monitored and regulated. Experience has shown that in actual practice these provisions are not consistently implemented over a long period of time. Chemical inhibition of corrosion by calcium nitrate is being developed as a means of slowing corrosion if caught at an early age. It is most effective in permeable concrete. However, marine and offshore concrete should be, and usually is, designed and constructed for impermeability; hence this system has limited applicability.

Now let us try to understand about the importance of foundation repairs of offshore structures. The foundations of offshore structures are an integral part of the system that resist the environmental forces and support the operational loads. Hence, they must be maintained in their design condition.

Scour around jacket-type structures is of two types. Of is an areal scour, which lowers the seafloor in a dish-like shape. The remedy here is to dump or chute in rock which will have two characteristics:

It is small enough so that it will not work its way down into the soil.

It is large enough so that it will not scour under current and wave-induced forces.

A permeable material is required in order to prevent trapping wave-induced pore pressures which can promote scour.

The above requirements can be met with a two-layered system, consisting of small rocks or gravel, the other large rock, but this is very difficult and costly to place around and between the legs of a jacket at great depths. A single blended placement of rock is more effective, requiring more quantity of material but enabling placement in one operation. Another effective scheme for scour prevention is to replace the small rock by a filter fabric and then cover it with large rock. The filter fabric can best be assembled in two layers, sewn together. One layer is a fine mesh to prevent sand migration; the other is a coarse mesh of heavy polypropylene, reinforced with stainless steel wire.

Mattresses can be made up of reinforced filter fabric with concrete blocks attached. They can then be lowered into place, using a steel frame. Mattresses filled with fine and coarse gravel have also been used. Still another system employs an integral filter fabric to which multiple closely spaced neoprene bags are affixed. After placement, the bags are pumped full of grout.

The other type of scour is localized erosion around individual legs. This is often accompanied by gaps forming around the head of the pile as it deflects under lateral loads. Here the placement of a mound of small rock (*e.g.* pea gravel) has proved to be very effective in the clays of the Gulf of Mexico.

A gas blowout under a platform can lead not only to cratering but also to a general loss of support for the piles. To restore the capacity and safety of the foundation after such an event usually involves a major effort. First, the crater can be filled with small rock or coarse sand placed through a tremie tube. Then both the newly placed material and the adjoining sediments most be consolidated, since the entire zone will probably have been significantly loosened by the escaping gas. The consolidation has to be approached with great care, since the resultant settlement of the upper soils may lead to distortion of the structure. For that reason, densification by shock (dynamic compaction) or vibration may not be suitable. Placement of an extensive blanket of rock may be safer, with careful monitoring of both soils and structure at frequent intervals.

When, due to accident or blowout, the pile capacity has been reduced, insert piles can be installed through the primary piles and joined to them by grouting. These insert piles are extended beyond the tips of the primary piles by drilling to obtain additional bearing and friction load transfer for axial loads. Insert piles, properly grouted, will also stiffen the piles and hence improves their capacity to resists lateral loads. As an alternative to a drilled and grouted inserts pile, a belled footing may be constructed using a belling tool. This solution may be preferable in stratified soils or where skin friction transfer is uncertain.

Because the overhang of the deck will probably limit the accessibility, requiring the insert pile to be assembled in short lengths and preventing the use of a hammer, a riser can be installed and reverse circulation drilling along with a slurry, is employed. Once the hole is drilled, the insert pile can be lowered to position, using either welded joints or mechanical connectors. Then grout can be injected into the annulus, bonding the insert pile to the soil and to the primary pile.

The above assumes that geotechnical investigations show adequate soil at the greater depths.

On two of the early concrete structures in the North Sea, after the platform was installed, drilling was commenced in order to install the conductors. As the drills penetrated the temporary closure of the concrete base slab, with the water level in the drilling shaft being lower than sea level, substantial quantities of foundation sand rushed up into the drilling shaft, leaving a void of several hundred cubic meters below. Some "piping" even led in from the outside perimeter due to the difference in heads. To repair and fill this void, an underbase grout mix having a low heat of hydration was flowed in under the slab, working in several stages to ensure complete fill. The zone on the periphery which had been disturbed by the "piping" was filled with small rock. Subsequent behavior of the platform, over a period of 25 years, has been entirely satisfactory.

To prevent this adverse phenomenon on future platforms, water levels in the drilling shafts are required to be equalized, especially during conductor installation. Steel skirts, surrounding and isolating the drilling compartments are built into the base slab design.

The Process of Increasing the Capacity of Existing Piles

Increasing the capacity of existing piles to sustain increased axial loads is a problem which has been encountered a number of times in cases where calcareous sands have been encountered and where a reevaluation of their performance under dynamic uplift has led to the decision to increase their axial capacity. It also occurs where a platform must carry heavier loads than those for which it was originally designed.

The most common decision is to install insert piles. Since the overhang of the platform usually limits the clear working space to 10 m or less, the insert pills will have to be installed by drilling rather than driving. Therefore, a work deck is installed, upon which a short drill rig can operate. Then a riser pipe must be run down to the top of the existing primary pile and sealed to it, sufficiently tight to enable an 8- to 10-m positive head to be maintained within the riser. Flange or coupling joints are used for the riser. Then the drill string is assembled and a hole drilled for the insert pile to the required depth.

The insert pile must be made up. Welded connections are standard, but a great many are required due to the short length of each segment that can be fitted in under the deck. The welding has to be carried out under conditions of spray and wind; hence, protection is required. Nondestructive testing, usually X-ray, takes time and clear access. The alternative is to use a high-strength tapered-screw mechanical coupling. Such fittings should be qualified by tests under dynamic and cyclic loading.

A grout pipe is run with the insert pile. An inflatable packer is provided at the tip to prevent grout from filling the insert pile. Alternatively, a float shoe may be installed. Cement grout, sand-cement grout, or fine concrete is then placed, until instrumentation shows that the annulus is full.

Shear transfer to the insert pile can be enhanced by welding on shear rings. *See* Figure 8.6. Means must be provided to prevent the insert pile from floating in the denser grout since the insert pile itself will usually not be filled with concrete, at least not simultaneously. It may be wedged to lock it in place or weighted by filling with barite drilling mud.

To save the time and cost of coming out with the drill string and then setting the insert pile, it may sometimes be expedient and adequate just to grout the drill string in place in the drilled hole using an expendable bit. This was done, for example, on several piles for strengthening of the Kingfish platforms in the Bass Strait, Australia.

Another solution is to utilize the end-bearing capacity of an existing pile, since this will have often been disregarded in the initial calculations. To do this, the material inside the existing pile must first be removed. On the same Kingfish platform referred to above, elongated holes were cut into each jacket leg, above water, for access to the pin pile. A concentric group of pipes was made up, using segments only 4 m long. These sections were progressively made up

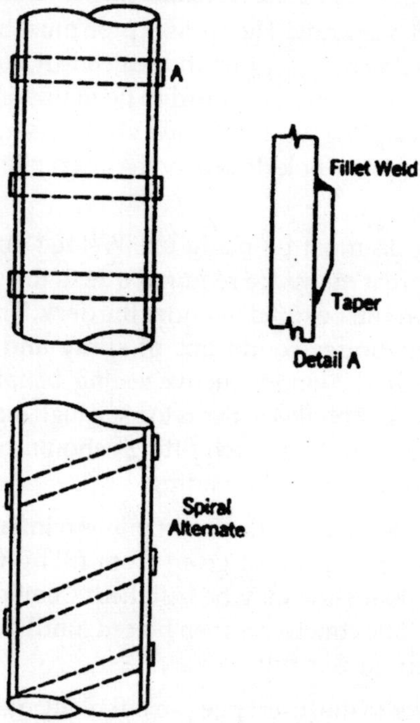

Figure 8.6: Shear Keys on Insert Piles

and lowered down the existing pile. Obstructions such as neoprene pile closures had to be fished out. Using the jet plus airlift, the plug was cleaned to within a few meters of the bottom. Then a sand-cement grout was injected to form a 10-m-long plug. Then an illmenite (iron ore) slurry was fed into the pile to give its added weight to resist uplift.

Pressure grouting around the tip and walls of an existing pile have been often proposed. Holes can be perforated in the walls of the pile by controlled explosive casing-perforator devices. Grout of very high fluidity and low surface tension can then be followed into the piles, to exit from the ports and bind with the soils. Special admixtures are available to help penetrate sandy soils. The problems are many, however. The sands, being saturated, are relatively impermeable, since any grout flowing into them must push water out. Grout always follows the path of least resistance, trying to find an exit to the sea. Thus while on paper we can draw symmetrical

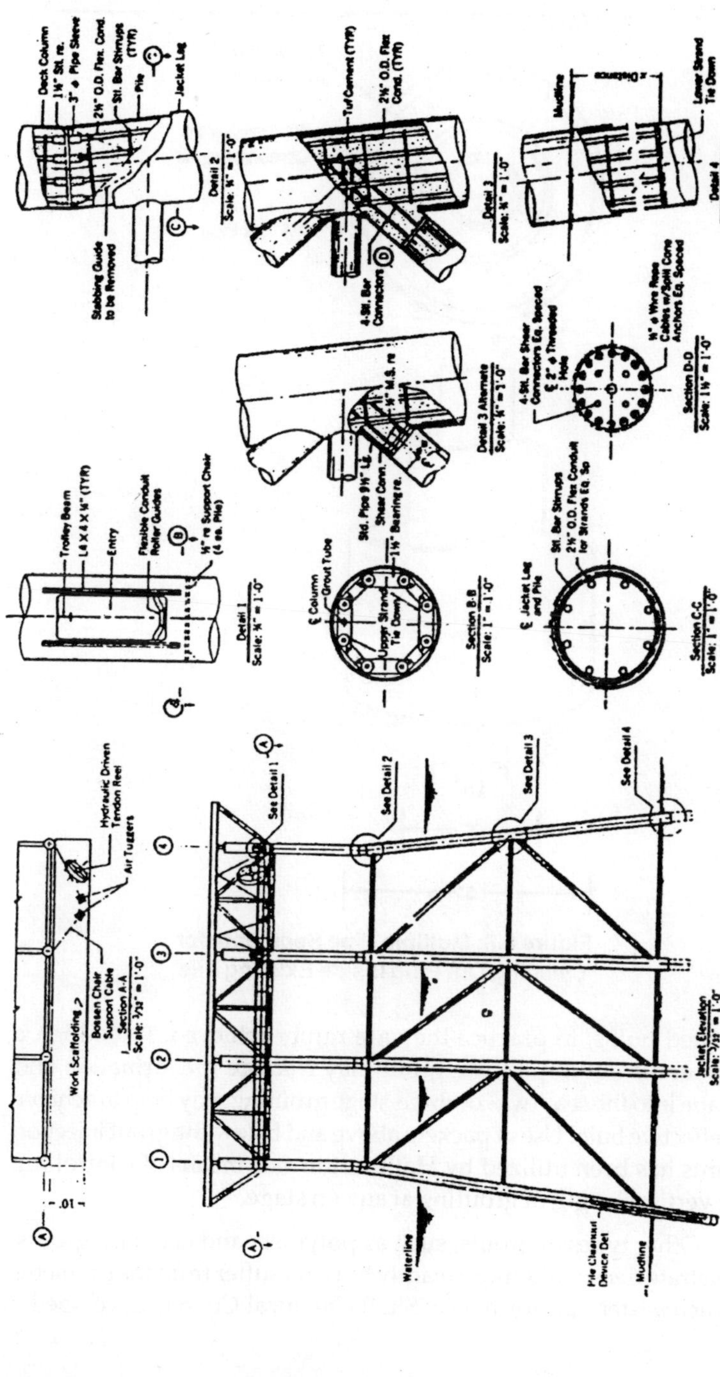

Figure 8.7: Platform Strengthening

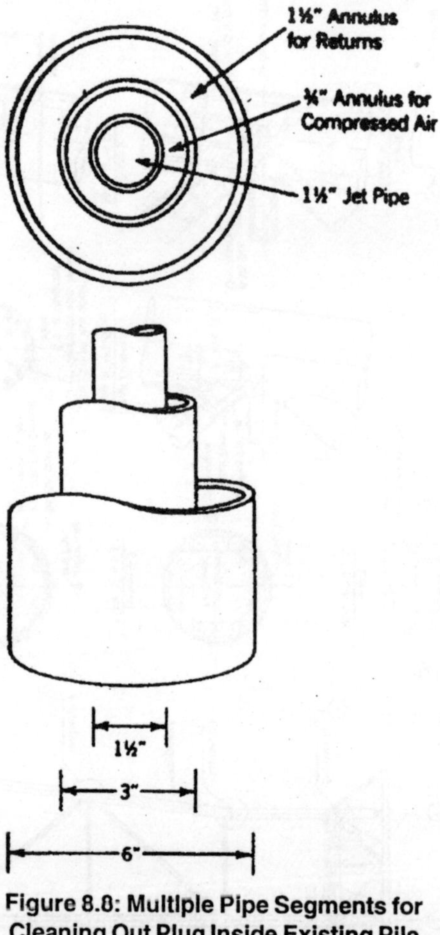

Figure 8.8: Multiple Pipe Segments for Cleaning Out Plug Inside Existing Pile

grouted bulbs, in practice they are rarely achieved. If too high a pressure is developed, the grout may fracture the formation and escape into the sea. Two- or three-stage grouting may help to achieve an effective bulb. Use of packers above and below the grout injection points has been utilized by Halliburton as a means of controlling the vertical length of grouting at any on stage.

Other types of grouts, such as polymers and chemical grouts, penetrate dense soils more readily but also suffer from the problems of inconsistent performance. Shell Chemical Co. has developed a low-viscosity epoxy material. Eposand, which penetrates sands of low permeability. This requires initial injection of fresh water, followed by injection of glycol or alcohol, after which the epoxy can be injected. Hence, it is time-consuming and expensive. However, it has proved effective in calcareous sands of low permeability.

The ballasting or weighting of a pile to give greater uplift capacity is similar to the weighting of a table leg. Earlier, use illmenite slurry was reported. Barites added to the drilling mud are expensive but effective and can utilize the drilling and mud facilities already on the site. Finally, magnetite iron ore graded for maximum density can be used.

Depending on the character of soil, drilled and grouted insert piles may not be suitable or may be excessive in length. Where good bearing strata exists near the tip of the existing pile, a belled pile footing may be constructed in much the same manner as the drilled and grouted insert pile.

Belled footings were constructed under the main pin piles of the Ekofisk drilling platform in the North Sea. A riser was run, and then the drill string. After drilling to the desired depth (T.D.) the belling tool was operated, gradually belling out a truncated cone. Reverse circulation was used, with airlift assist. The air injection was well above the bell, up in the primary pile, so as not to reduce fluid density in the bell and cause caving. When the belling was completed, spot mudding may be used, if necessary, to hold the bell open, although the internal head in the riser, being in a meter or two higher than sea level, helped to stabilize the bell by causing all flow to be outward. Either a bentonite mud pre-converted to calcium bentonite or a polymer mud should be used to be compatible with the cement.

Next the bell is concreted. A concrete containing small aggregate, up to 10 mm maximum, is preferred to straight cement grout because of higher tensile strength, lower heat of hydration, and less tendency for a bitter mode of failure. In any case, the cementitious materials should have low heat of hydration: a 70-30 blast furnace slag-Portland cement mix is suggested. Alternatively, a cement with 50 per cent pozzolan replacement may be used. The mix may be precooled.

The tremie process using gravity flow is the preferred method. Pumping has been employed but may develop excessive laitance due to disturbance under the high rates of flow and surge. The tremie method appears to give better control, especially if the formation surrounding the bell is sensitive to the hydraulic fracture.

Shear transfer from the insert pile to the bell has to take place over a relatively short length, and hence special shear transfer devices are usually required, such as reinforcing bars grouped in bundles or shear keys on both inside and outside of the insert pile. End bearing plates may also be used, providing direct load transfer from the insert pile to the bell. A float shoe or end plate may be provided at the tip of the pile to prevent concrete from filling the insert pile. The insert pile must be weighted or secured to the main pile to prevent it from floating. A secondary system may be necessary for grouting the annulus between insert and primary pile, in order to ensure complete filling of this critical zone.

On the North Rankin platform, off the Northwest Shelf of Australia, it had been discovered during construction that the piles had little if any frictional resistance in the calcareous sand, despite a penetration of 120 m. the platform had already been completed and was operating. Hence, all remedial operations had to be carried out from underneath the deck. The decision was made to conduct belled footings under each of four primary piles in each of the corners, a total of 16 piles. Since these were skirt piles, with heads some 80 m below water, a casing was run, through which the drill string could operate.

The drill then drilled a straight-sided hole extending about 10 m below the tip of the driven pile and 1 m below the design elevation of the bottom of the bell. The proprietary epoxy Epoxand was injected to stabilize the surrounding sands during belling.

The belling tools now expanded slowly to enlarge the bell to a bottom width of 4.5 m. Then the insert pile was placed. Based on tests, the end plate on the insert pile, with a slope of 7°, was founds to be sufficiently sloped to ensure full filling on the underside, with no trapped air or bleed water. Then a 3½ in. (88mm) tremie pipe was installed. Concrete mix was 8 mm sea gravel, sand, blast furnace slag cement, with fly ash and silica fume. Admixtures used were plasticizing aid retarding (SIKA Plastiment) and a superplasticizer. Concrete was precooled to 5°C by injection of liquid nitrogen. The ambient temperature of air, water, and soil was 38°C. The tremie pipe was precooled by cold water. Then, using a pig, the tremie concrete mix was placed.

The main pile and consequently the casing and insert pile all slopped at 7° from vertical. This turned out to be advantageous in the concrete placement, in that entrapped air could escape quite readily from the tremie pipe. Concrete was transported from the mixer to a hopper through an insulated pipeline, thence by gravity flow down the tremie pipe.

Cores taken subsequently indicated that the design strengths of 45 MPa compression, 7 MPa tension, were met and exceeded.

The Importance of Modules

In recent years, almost all topside facilities have been first fabricated into modules and then transported by barge and set on the platform by an offshore derrick barge. The capacity of offshore derrick barges has steadily grown to where 1200-ton modules are commonplace and individual lifts of 4000 to 11,000 tons and more have been made.

The purpose of using large modules is to enable more of the fit-up and testing to be completed at the shore site. This not only has the advantage of enabling the work to be done under optimal conditions, but disperses the work so that it can be accomplished concurrently with other modules and other structural work.

The modules are set onto the module support frame, which is a skeletonized deck structure. Some will be set onto skid beams and skidded and jacked to final position; others may be set directly. The modules must be structurally adequate in themselves both for the temporary loads imposed during transport and installation and for the permanent loads due to the operations and environment. The structural support frame for each module must first support the vessels and the piping within it and then transfer the forces developed by dead and live loads and environmental loads to other modules or the module support frame.

Lifting of such extreme loads must follow the general principles of heavy offshore lifts outlined previously and must be thoroughly engineered for all stages of the operation. Picking points and padeyes must transfer the forces to the slings. The slings, with their angles in three-dimensional space, must in turn transfer the loads to the hook.

Where more than one crane will be involved in the lift, the interaction of loads between the cranes must be considered, including the effect of tolerances in boom position sea-induced motion, and the change in the derrick barges water planes as the load comes onto them.

When picking a module with multiple pickup points, deflections must be carefully controlled so as not to distort the equipment and piping. Very elaborate rigging systems often result. The frame may have to be stiffened.

Picking loads are dynamic; adequate allowance must be made for dynamic amplification in lifting force, as well as in lateral swing. This latter can be greatly reduced by power-controlled tag lines. Low-temperature effects, possibly causing embrittlement under impact loads, need to be addressed and suitable steels and welding procedure adopted.

Many modern heavy lifts of modules are assisted by onboard computers monitoring the loads, the radii, and the position of the booms. Some cranes barges are equipped with boom tip motion sensors and onboard computer systems to determine the best headings and boom angles to minimize boom-tip motion.

Modules are usually loaded onto a barge at a shipyard or shore base by skidding out, much as a jacket is loaded out. Dimensions are smaller and total weight much less, but loads may be more concentrated. Alternatively, they are loaded by transporters. The modules must then be properly tied down for sea.

Engineered slings are pre-attached to each module so that all that remains to be done as the lift commences is to raise each sling up over the hook by means of the crane while line. Meanwhile, the tie-downs are cut loose.

When sea conditions appear favorable, the module is lifted clear of the barge, slowly rotated to position, and set in its place. Auxiliary means such as powered tag lines on the deck of the platform, tapered guides, and fenders are used to help seat the module in correct position. The module needs to be set down smoothly and quickly so as not to expose the operation to higher waves or low-cycle fatigue. It is desirable to incorporate tolerance in initial positioning into the structural design. Once the unit is set, jacks can then move it to final exact position. Jacking points should be provided in the module frame.

The problem of overhaul during set-down–that is, of getting rid of a load from the hook when there are 24 or more parts of line in the hoist blocks–is a difficult one. A free overhaul clutch for the crane hoist is the primary solution. In some cases, ballast may be transferred to the stern of the crane barge as soon as the load touches down. The object is to prevent the load from being inadvertently lifted back up as a subsequent wave raises the derrick barge's stern before there is enough slack in the falls.

Some rather spectacular lifts of modules have been made by the use of two or even three crane barges working in concert. The three Statfford A quarters modules were too high (40 m) to enable them to be lifted by a single crane. The weight of each, about 1000 tons, was not too unusual, but the height and profile required that three crane barges be used. These were moored together with all deck winch controls and dynamic-positioning thruster controls at one control location. The three barges picked up the module at the dock, transported it to the concrete gravity platform moored in the fjord, and re-positioned the barges while carrying the load. At the final lifting site, the barges were moored to structure. Because of the short length of lines, nylon rope was used to have some elastic stretch to accommodate surge as the quarters modules were then raised and set on skid beams mounted on the deck frame. Each module was then skidded sideways to its final position. The pick had to be engineered with extreme care, since the loads exceeded the capacity of each crane barge by itself.

On a subsequent platform, similarly high and heavy modules were set by a semisubmersible crane barge. Since this work was carried out in a fjord, the semisubmersible was not selected for minimal response to seas but rather for its extreme height when deballasted to ride on its pontoons. Now it was able to lift the module over the deck structure and to set the module directly in place. Up on the platform deck, each module is then welded to the module support frame.

Now let us learn about the hookup of the modules. The hookup of the modules and their subsequent testing is highly demanding in terms of manpower and support. It delays the start of production of oil or gas and hence adversely affects cash flow. In recent years, the complexities of hookup have led to overruns in cost and time of 100 per cent or more. To reduce these, the first step is to use larger and fewer modules, that is, more self-contained modules. A second step is to space the modules apart by 1 m or so to allow a crawl space for access for interconnection. A third step is to use flexible connections to the extent permissible for the high operating pressures in the pipeline connections.

Careful control in tolerance of all interconnecting points at the time of module fabrication is essential. Templates and pre-matching may be used to ensure compatibility.

If the hookup work is supported by a semisubmersible derrick or "floatel," a suitable gangplank or walkway is required. This must have roller to accommodate surge of the semisubmersible. It must be supported, for example, by cantilevering, so that it will not fall even if the barge drags an anchor or parts a mooring line. Sophisticated articulation is often provided to compensate for much of the barge's movements.

Fire protection during hookup is critical and must take precedence over actual work. This requires the early installation of the fire pump casting, the submersible pump, and headers around the platform decks. Until the platform system is fully established, fire hoses must be led over from the tending semisubmersible or derrick barge. A fire alarm system must be hooked up.

Life safety must be ensured; this means that life jackets and safety lines must be employed initially, that lifesaving capsules must be placed with the initial modules, and that a patrol boat must be on duty to pick up anyone who falls overboard.

Other services and systems on board the platform must be activated at an early date. These include the generators, both primary and emergency, with their diesel fuel supply tank, and lighting systems for night work. A freshwater system must be established for portable water supply and wash down. Compressed-air systems are required both for instrument air and for utilities and tools.

Radio communication to shore and boats must be established, as well as public address systems throughout the platform. A smoke-detection system is needed in the quarters modules, as well as sprinkler and fire alarm systems. At the helideck, a foam extinguisher system must be installed. The rope landing nets must be installed and the landing lights activated. Ventilation systems must be activated in the quarters and within the facilities area.

Because, during the hookup stage, welding is the principal item of work, welding generators must be installed and cables led around the deck. Heated welding rod storage needs to be provided.

In addition to the permanent platform cranes and hoists, temporary hoists and powered winches will be required during hookup. Adequate lighting must be provided for night work. Temporary shops and offices are needed. An X-ray lab is required. There must be an electrical shop, an instrument shop, and a general tool shop and a warehouse for bolts and pipe flanges. A paint shop is required, with its separate fire-extinguishing system. Finally, a first-aid room must be available for emergency treatment of injured personnel, *e.g.* for burns and eye injuries.

It is obvious from this long summary listing that each module should be equipped as far as possible with the items needed to complete its hookup. Beyond that, very thorough and detailed planning is needed by engineers, crafts supervisors, and their forepeople to insure that all needed supplies and materials arrive with the modules and do not requires the separate lifts. However, these separate items cannot be just loosely stored in the module must be properly boxed and secured so that they cannot be displaced during the lift. Further, their weight needs to be computed and added to the calculated lift weight. The module lift weight then becomes the sum of the equipment, piping, cables, module frame, lifting gear, including slings, and the tools and supplies stored on board.

Vertical access has proved to be a significant problem, due to the heights involved from waterline to deck, and the large number of personnel. In addition to well-constructed stairways, construction elevators are usually installed.

For very heavy offshore lifts, a large semisubmersible derrick barge with two cranes at the storm has used both of its cranes in concert, holding one dead astern while rotating the other as necessary to position the load on the platform. Combined capacities of 10,000 to 13,000 tons are thus available. For such operations, freedom from long-period swells is necessary. Work is carefully scheduled for the weather window. The derrick barge may sit several days awaiting suitable seas in which to make the lift. Of course, the large semisubmersible barge does not react to short-period wind waves of moderate height, but it is sensitive to longer-period wave energy such as that from remote storms.

A complete deck or very large section of the deck is constructed on a quay wall or in the prefabrication and assembly yard adjacent to the quay wall. It is then either skidded onto a large transport barge or walked out by multiwheeled transporter, in which each of many wheels has its load equalized by hydraulic jacks. The transfer onto the barge has to either be with the barge grounded at the quay wall or the barge has to be equipped with a variable ballast system enabling it to adjust to change in load as the module moves onto the barge. The barge then is towed to the site and moored at the stern of the crane barge. The cranes lift the module above deck level, the barge is towed clear, and the barge moves into the platform. It then lowers the module onto the deck, using a guide system.

With all such monstrous deck modules, the deflections and stresses due to dead load during picking need to be carefully computed and special means taken to accommodate the dimensional changes in support locations that occurs as the module is picked and then set. The DNV Rules contain an appendix for Heavy Lift Operations, which is applicable to these operations.

Chapter 9

The Submarine Pipelines

Introduction

Almost all steel pipelines have been joined by full-penetration welds, especially in the petroleum industry, where pressures typically run 1500 psi or more and leakage of oil or gas is unacceptable. Consideration is being given, however, to the use of mechanical joints, for example, joints similar to those used with well casing. Development work continues on explosively and hydraulically expanded connections. In a few cases, flanged connections are used, but these are for lower pressures.

Since most submarine pipelines are installed empty, they are subjected during installation to high hydrostatic pressure, along with whatever bending may be taking place. They are laid under axial tension; thus buckling under combined loading becomes a principal design consideration. Tolerances are consequently of great importance, out-of-roundness and wall thickness being the most critical.

The steel is protected from external corrosion by coatings such as bitumastic or epoxy, supplemented by cathodic protection, usually sacrificial anodes. Internally, the line may be uncoated if it is to be in petroleum service, or it may be internally coated with epoxy, polyurethane, or polyethylene or cement-lined when it will carry seawater or corrosive substances. The external coating may be further protected from abrasion by concrete or fiberglass wrapping.

To give stability to the line when in service, especially those lines which must be emptied at some stage of their life or which carry a low-density material like gas, the line must have net negative buoyancy. This is usually supplied by concrete weight coating (which can also serves to protect the anticorrosion coatings) or by increasing the wall thickness of steel. A number of pipelines have experienced "floating up" off the seafloor due to spalling and shedding of their concrete coat. This indicates that the reinforcing mesh may have been underdesigned or that the pipelines may have been subjected to excessive overstress during installation.

The latter is within the purview of this book. It appears that most such cases of damage have occurred during the more severe sea states when the pipe laying barge was subjected to severe dynamic surge. If the coating was not only cracked but delaminated from the pipe, then transient pore pressure under the storm waves break the coating off in progressive failure. This type of failure has occurred a number of times during pipe-pulling operations. The most obvious solution is to increase the amount of circumferential reinforcing in the coating. Since the coated pipe is usually furnished by the oil company, this obviously presents a contractual problem to pipeline installation contractors. Nevertheless, contractors may often find it in their best interests to verify the amount of circumferential reinforcing and, where necessary, recommend increased reinforcement.

Pipelines are basically designed to lie on the seafloor or in a trench in the seafloor, with more or less continuous support. However, unsupported spans may occur in rough, rocky seafloors or where the sands move under the action of currents and waves. The designer will have set limits on the unsupported span lengths, which the contractor must not exceed; this may require either prior seafloor leveling or post-installation support.

Lines are buried beneath the seafloor in many areas of the world to protect them from fishing trawl boards, from dragging anchors, and from fatigue due to oscillation in a current. Sometimes the trenches are backfilled with soil or covered with rock but in most cases natural sedimentation is counted on to fill the trench.

As noted earlier, the pipeline usually sees its most severe stresses during installation; thus very close integration is required between the designer and the installation contractor. The designer needs to be aware of and address the needs of the contractor during installation. The contractor conversely must be aware of the limitations and constraints imposed by the installation procedures, taking into account the sea state (waves and current), the varying water depths, and the varying seafloor.

In addition, all parties must be cognizant of other pipelines and facilities in the area, recognizing the tolerances in location both of the previously laid lines and facilities, and the tolerances which are inherent in the contractor's procedure.

Submarine pipelines are typically laid in a "corridor" whose centerline and width are given by the client and shown on the approved permit. The installation contractor must have an adequate survey system to enable the contractor to comply. This system is usually an electronic positioning system or real-time differential GPS but may include lasers, ranges, and preset spar buoys.

The installer must verify to the satisfaction of the client and the regulatory body that the line has been satisfactory installed. Externally, this is done by side-scan sonar and ROVs, using video or acoustic imaging. Internally, the line is pigged and then tested with hydrostatic pressure to a pressure in excess of the design pressure.

A pipeline "pig" is a short cylinder, of slightly less diameter than the pipeline, with several sets of squeegee wipers. When the pig is entered in the pipeline and excess pressure is applied, it travels along the pipeline. The diameter of the pig and its length verify that there is no dent, crimp, or buckle more than the smaller annular space. The squeegees hold the pressure so that the pig will move. The pig is usually equipped with an acoustic transponder or radioactive marker so that if it does get stuck, its position can be determined. Guidance for the design and installation of submarine pipelines is given in the DNV *Rules for Submarine Pipeline Systems.*

The Offshore Lay Barge

The offshore lay barge has grown up from the specially modified cargo barge of the 1950s to become one of the most sophisticated, efficient, and expensive vessels in the world. Lay barges are often characterized as first-, second-, third-, and fourth-generation to donate major quantum jumps that have been made in the extending the ability to lay lines in deep water, with current achievements being the successful installation in the depths over 600 m (2000 ft)

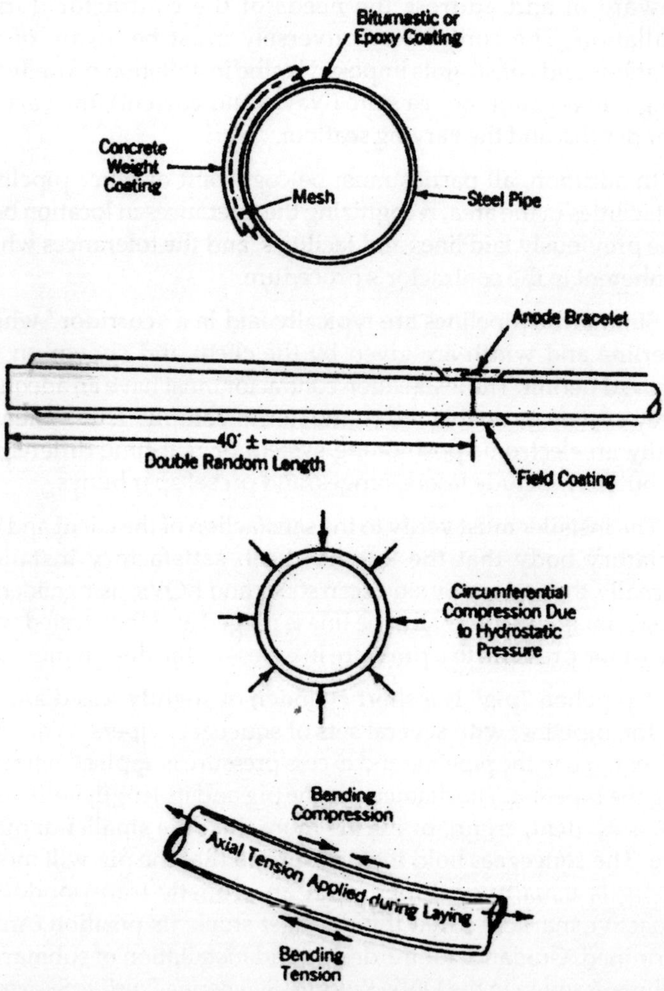

Figure 9.1: Typical Steel Submarine Pipeline, Showing Stresses Incurred During Installation

across the Strait of Sicily and in such adverse environments as the North Sea.

First-generation lay barges have a conventional barge hull, with the pipe-laying assembly mounted on one side. The stinger is hinged but rigid. Second-generation lay barges have a semisubmersible hull, with the pipe laying assembly on one side and an articulated stinger.

Third-generation lay barges lay pipe on the centerline, over a fixed cantilevered stinger. First-, second-, and third-generation barges all have deck engines and mooring lines. Fourth-generation lay barges use dynamic thrusters and a fixed cantilevered stinger. They are usually equipped for both S-lay and J-lay operations. These arbitrary distinctions are descriptive of the rapid advances which have been made in pipe laying technology.

The lay barge is positioned on its anchors, 8 to 12 in number, holding it aligned with the pipeline route, with a "crab" or slight orientation angle as needed to accommodate the effects of the current. Its position is determined by an electronic positioning system or GPS, augmented by laser in some cases. Its orientation is by gyroscope.

The anchors will be progressively moved forward as the laying take place, usually in 500 to 600 m jumps. One anchor-handling boat on the starboard side will move each anchor ahead in succession; another anchor-handling boat will move each of the port anchors ahead in succession.

Typically, the anchor-handling boat maneuvers close to the anchor buoy to enable the deckhand to hook an eye in the end of the pendent. The deckhand attaches a wire line from the deck engine of the tug, which either pulls the buoy abroad or pulls the pendant through the buoy, thus lifting the anchor clear of the bottom 5 m or so. The boat then runs forward, setting the anchor as directed in its new position, releasing the buoy. The boat turns outboard and goes back for the next anchor in the cycle.

The new position of the anchor is given by voice radio command from the control house, which is based on radar, gyro, and the reading on the remote mooring line length counters, reading the line length paid out by the winch.

The proper paying out and taking in of each mooring line on the winch drum is monitored by video in the control house to ensure against crossed lines on the drum or fouling of the line.

From a supply boat or barge alongside the port side, the crawler crane on the lay barge snags (picks) one pipe length (12 m) at a time, turns, and sets it in storage. From storage, the crane picks a pipe length and sets it on the end-O conveyor which moves it to the transverse conveyor at the bow. This conveyor feeds it onto the lineup

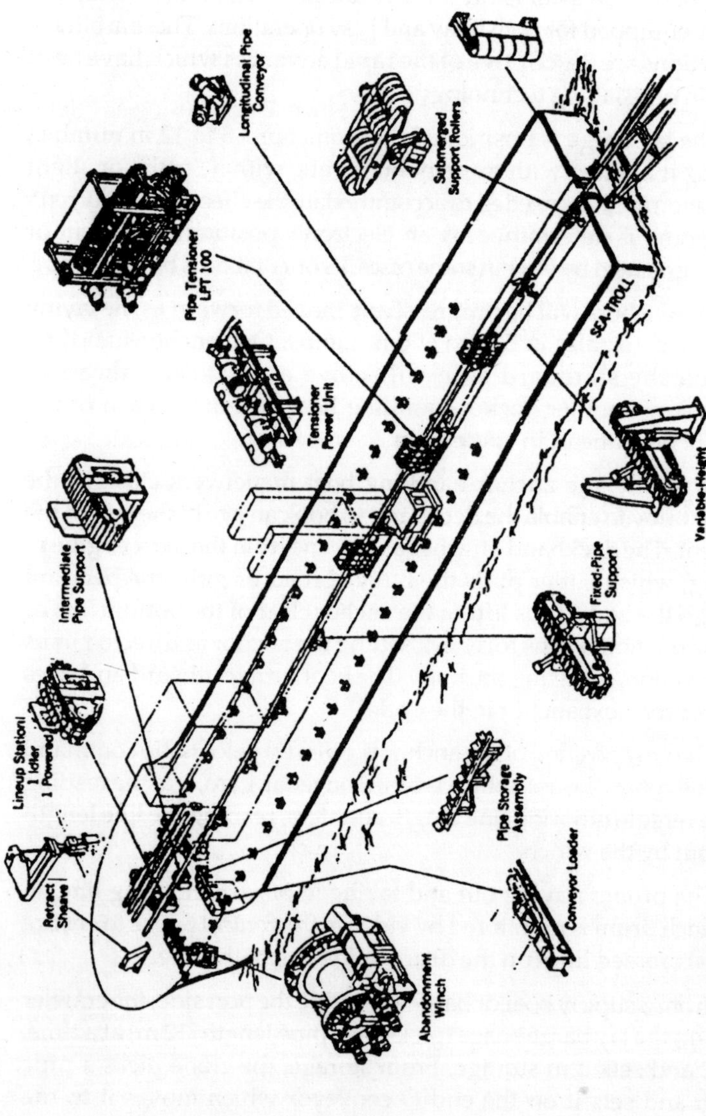

Figure 9.2: Layout of Equipment on Ship-shape Pipe Laying Vessel

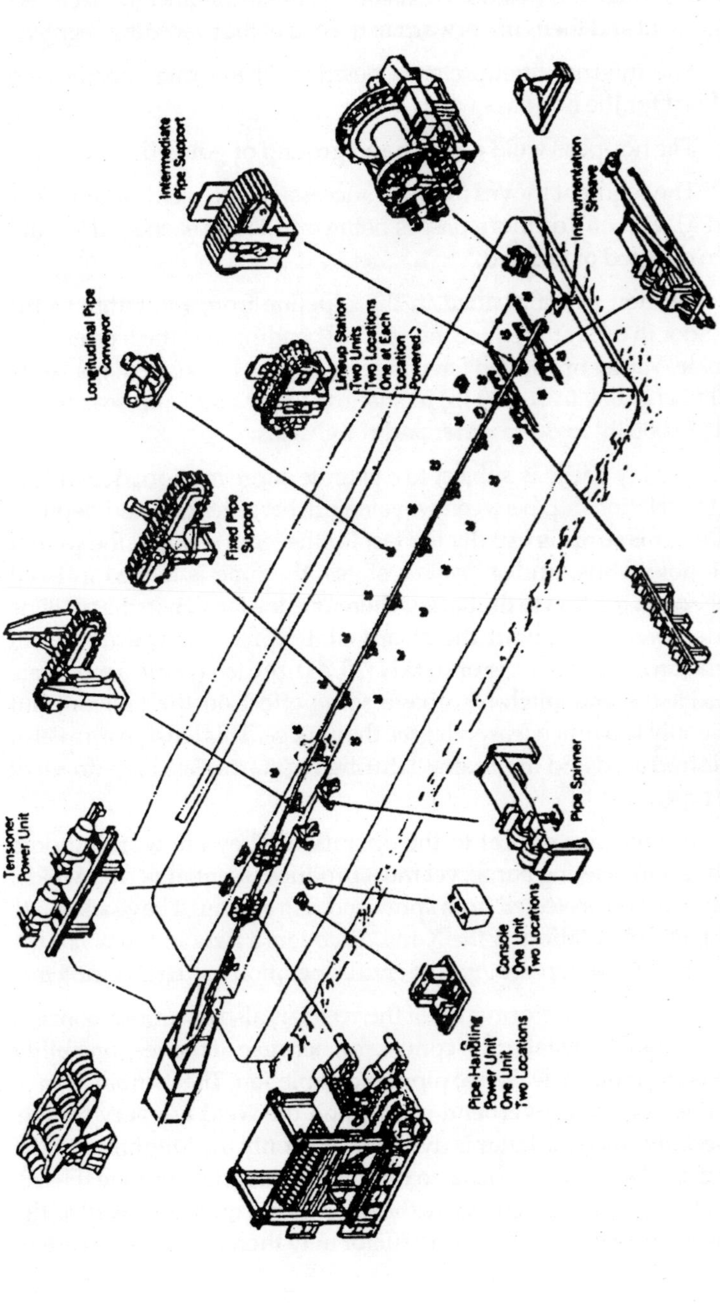

Figure 9.3: Equipment on Modern Pipe-laying Barge

station, where it is positioned, usually semi-automatically, in correct alignment and then run forward to the end of the preceding segment.

The internal lineup clamp positions it in exact spacing and holds it for the hot-pass weld.

The hot-pass weld is made and ground or gouged.

The segment moves forward successively to weld stations 2, 3, and 4, with one or more passes being applied at each station, and then chipped or gouged.

Tension is maintained in the pipeline from the barge to the seafloor in order to reduce the vertical bending and the tendency to buckle. Values of applied tension range from a low of perhaps 100 to 150 kN (20,000 to 30,000 lb) in shallow water and calm seas to 300 kN (70,000 lb) in deep water and rough seas.

The lay barge is subject to dynamic surge motion, depending on the relationship between wavelength, barge length, and depth of water. This surge is usually too fast for the tensioner and the welder to follow. Thus, under low sea states, the pipe is locked in fixed position in relation to the barge. Therefore, the tension in the pipeline varies cyclically about the steady-state force. Typical ranges of variation are of the order of 100 kN (20,000 lb) each way in a moderate sea. Heave and pitch also have some effect on the tension, but generally to a much leaser degree than surge. This tension must also be introduced and maintained during the startup and lay-down of the pipe.

Welders are critical to the operation. They are working on a rolling and heaving barge, yet must produce essential perfect welds. They must be protected from spray and rain and must have adequate light and ventilation. If the X-ray discovers a flaw in the weld, the resultant cutout repair stops the entire operation until it is completed.

The actual performance of the welds is also of serious concern to the pipeline installation contractor, because of the responsibility to ensure a sound, leak-free pipe on completion. The combination of axial tension and overbend stresses on the weld are very severe, especially since the latter is dynamic. Not only the toughness of the weld itself is involved, but also that of the heat-affected zone (HAZ), which in turn is influenced by the parent steel quality as well as the welding procedures. The constructor may therefore find it prudent to test the pipe steel and welding procedures under dynamic tension load prior to finalizing procedures.

In a typical offshore operation, the barge will move one pipe length every 15 min. On the most modern third-generation barges, using advanced welding techniques and double- or triple-ended pipe joints, rates of a mile per day are achieved. This means that all the work must be completed at each station within the same time frame. This translates to 100 or more 12 m lengths per 24 hour day. These performances have been exceeded by top-notch crews on good days, even with manual welding.

Stresses in the pipe in the laying operation are controlled not only by axial tension but also by the net submerged weight of the pipe. This latter is the difference between two large numbers, the one being the air weight of the pipe, the other being the buoyancy due to the displaced volume. The major variable is the thickness of mortar coating, which affects both air weight and displacement, both not equally.

In a typical case, a pipe may have an air weight of 15 kN/m (1000 lb/ft) and a displacement of 14.3 kN/m (950 lb/ft) leaving a net (buoyant) weight of 0.7 kN/m (50 lb/ft). If the coating increases the weight by 5 per cent, the displacement may increase only by 2 per cent. These numbers may sound small, but they develop an increase in the buoyant weight of 0.45 kN/m or a 60 per cent increase in the force causing the bending. Thus while weight control is normally not as critical with lay barge operations as it is with bottom pulls, it nevertheless is of great importance and must be monitored.

The pipe is generally furnished in double-random lengths, which normally 12 m (40 ft). Most sections will run 11.4 to 12.6 m. However, generalized pipe procurement specifications allow a few sections which vary widely from the norms, even as short as 3 m and long as 17 m. While this may be accommodated on the land pipelines, it is unworkable at sea. Such sections should be cut or sliced to the normative length of 12 m (40 ft) at the shore base, or else the procurement order should exclude these variances.

As previous noted, rates of progress with third-generation lay barge may reach 1 mile per day or more. This means that 100 or more sections must be loaded out each day from the shore base, transported to the site, and then unloaded to the deck of the barge.

This last is a critical operation when the seas are running high and may, along with anchor handling, be the controlling operation. The transfer at sea of the pipe is a typical case of operations involving two vessels alongside each other, of different characteristics, each responding in its own way to the seas, in each of six degrees of freedom. The relative positions in plan can usually be maintained in a moderate sea state by trying the transport barge alongside the lay barge, with suitable fendering, so that the major individual responses are limited to heave, roll, and pitch. In heavier sea states, barges can no longer be kept alongside, and so supply boats are used. By running a stern line from the boat and so keeping power on, a good skipper can hold the boat in reasonably close position, although now the skipper will develop some relative sway, surge, and yaw motions as well as heave, pitch, and roll.

The typical lay barge is restrained from lateral motion by the mooring lines which move it periodically one barge length ahead. These lines, while catenary in scope in deep water, are kept under tension by the winches. The lines tensions are measured by tension meters on the wire rope or on the winch drum or both. In the typical second-generation lay barge, tension may be 400 kN (80,000 lb) with a variance in a moderate sea of ±100 kN. This variance is due to the long-period sway plus surge built up by the waves, storing energy in the wire lines as the barge gradually moves to one extreme of its lateral range. The lines on the far side gradually become more taut, so that eventually the barge changes direction and starts its sway excursion to the other end of the range. The acceleration at the end of its excursion causes a shudder effect in the overall system, which translates into a severe horizontal whip of the stinger and of the pipe. The surge excursions causes cyclic bending in the pipe at the overbend and in high sea states can lead to low-cycle fatigue in the pipe.

The mooring lines must provide the horizontal restraint against wave drift, wind drift, and current drift. They also react against one another and especially must counter the tension on the pipe, which in effect is like a mooring line of relatively equal tension, leading directly astern. Balancing out the tensions in 8 to 12 mooring lines plus one pipeline is a complex problem, especially when these line forces are not steady but subject to the significant ranges introduced by the long-period excursions.

Typically, the tensions in the mooring lines are set so that under the maximum design surges, the force will not exceed 50 to 60 per cent of the guaranteed minimum breaking strength.

To offset the pipeline tension requires additional mooring line forces in the lines leading forward. The system must be balanced up, which is difficult enough with one positioning of the anchors, but which is rendered more complex due to the constant lifting and relocation of anchors. The system can be satisfactorily resolved by preparing calculations of typical and extreme positions for each permutation; it, of course, lends itself to the use of an on-board mini- or microcomputer, which can then solve for intermediate situations.

The most modern lay barges (fourth generation) use dynamic positioning by means of thrusters to maintain lateral position and heading. These are controlled by computer and connected to the GPS system. However, reaction lines to the pipe tension, usually two forward leading lines, are still required because of the large forces involved when lying by the S-curve method, especially in deep water for which high tension is required in the pipeline. Dynamic thrusters eliminate the long-period sway and consequent kick back of the barge, making it practical to continue welding operations in higher sea states.

The cost of pipe laying is directly related to the progress, since the cost per day is more or less the same whether any pipe is laid or not. The rate of progress has until recently been controlled by the time required for welding. There is a specific amount of weld metal which must be applied. Only two welders (one on each side) can work at any station. Therefore, the rate of progress depends on the number of stations. Typically, these are placed one pipe length (40 ft) apart. There is only room enough for a certain number of welding stations on a barge; therefore, the longer the barge, the greater the rate of progress. This explains why prior double-ending of the pipe does not speed the operation. Another means of accelerating the welding is by the use of microwire welding, but this is usually only acceptable in hot climates because of the dangers of cold lap at lower temperatures. The biggest jump in pipe-laying progress has come with the introduction of automatic welding of one type or another. Dual-torch automatic welding equipment, riding on a self-propelled carriage, can complete quality welds at a high rate, even in rough seas. The operation is fully computer controlled.

Second-generation lay barges are limited by the sea state. When the significant wave height exceeds about 8 ft (2.54 m), operations must shut down. The specific limit, of course, depends on the relative direction and period of the waves, as well as the barge length and width. The limiting item is usually control of surge and interaction between stinger, pipeline, and barge. The working limits can be increased by using a wider and longer barge, by using more powerful tensioners, and by using a fixed cantilever stinger.

When seas reach 10 to 12 ft (H_s = 3 to 3.5 m), then other constraints arise. Anchor-handling boats can no longer pick up the anchor buoys, although this limit has been extended by clever arrangements enabling the boat to run past the buoy and snag it rather than having to back down for the deckhand to make fast to the pendant. Pipe transfer from a barge alongside is no longer practicable with an H_s of 2 to 2.5 m, but a supply boat can be used to extend this operation to the 3- to 4 m range.

The barge motions in roll and long-period sway (snapback) become too severe in higher sea water, especially with a beam or quartering sea, and the welders are unable to produce quality welds. The pipe starts to jump out of the stinger, and there is danger of buckling the pipe. Even with dynamic positioning, the long-period surge causes severe variations in the pipeline tension and profile.

At this stage, a decision must be made whether to hold on or to initiate abandonment procedures. The major factor here is the weather prediction. If improvement is forecast within the next few hours, it may be practicable to hang on, maintaining tension. Another factor is weather or not the anchor will hold in the seafloor soils or any likely to drag; a dragging anchor will almost always lead to a buckle.

When abandonment is decided, a bull plug (cap) is welded onto the pipe. A line from the constant tension winch is attached. A buoy and pendent are also attached to the bull plug. It is a good precaution also to attach an acoustic pinger to the bull plug. The barge then moves ahead, paying out on the line, until the pipe is fully laying on the seafloor. The end of the constant-tension line is buoyed and run off. The barge can now pick up its anchors to move to a sheltered location or decide to ride the storm out at sea, on its anchors, but turned now to head into the sea.

When the storm ends, the barge moves back to location and resets anchors. While one hope to find the buoys, it is not unusual for them to have been torn away by the storm. That is when the acoustic pinger is needed.

The constant-tension line is now pulled on-board and the tension applied. The barge slowly moves astern, bringing the pipeline back up onto the stinger. A line from the crane may have to be hooked on (by diver) to help guide the line back onto the rolls of the stinger without fouling. Now the pipe is pulled on board, through the tensioner, until the bull plug reaches the lineup station; the bull plug is cut off and the pipe end re-beveled, and the laying operation can recommence.

An important consideration is that abandonment procedures are almost always carried out under extreme conditions, at or above working limits, whereas recovery operations will normally be carried out in good sea conditions.

The start of pipe laying also requires special procedures. Assuming the work will start at the platform, anchor is set at a point astern of the platform and the line brought up the stinger onto the barge, where it is welded to a bull plug on the leading end of the pipe. Tension is applied by the tensioner and the pipe-laying operation commences, with the lay barge moving progressively ahead and the completed pipeline being pulled off the stinger. The alignment is set as close as physically and safely possible to the platform so that, when the bull plug reaches the seafloor, it will lay alongside the legs of the platform.

Another method is to pull the end of the line off the lay barge. This can be done by running a line, attached to the pipe end, from the barge to a sheave at the preset anchor and back up to a winch on the barge. This enables the end of the line to be pulled to the exact position desired. The potential problems here are fouling of the line at the sheave (*e.g.* the sheave flips over, jamming the line) or fouling of the line in the platform bracing.

The platform jackets may have been prefitted with a J-tube and a witch on deck so that it can pull the line off the lay barge. While this system can only be used with small-size lines, up to about 300 mm (12 in.), it is very economical and eliminates the need for a separate riser operation. Deep-water platforms have a pre-installed riser and subsea connector, into which the pipeline is pulled.

If a large boat is available and the seas are calm, it can pull the line off the lay barge and lay the end down adjacent to the platform. The efficacy of this method depends on the bollard pull being at least equal to the required tension.

When the line terminates at a platform, it is customary to lay past the platform on one side and then follow abandonment procedures, laying the line down on the seafloor.

Earlier it was stated that the most serious problem in pipe laying is a wet buckle. In the case of a dry buckle–that is, where the line does not take on water–the pipe can be just pulled back up on board. In the case of a wet buckle, however, the pipeline has been flooded and cannot be brought back on board without creating continuing buckling. For this reason, at start-up, at least one pig was placed in a pig chamber at the start-up end, along with air fittings. If a wet buckle occurs, compressed air lines are connected and the pig run along the pipe to the point of buckle. This empties the line so that it can be recovered.

Actually, there is one even more serious case, known as a propagating buckle. This is the case where the ovaling the pipe at the point of initial buckle reduces the collapse strength below the resistance to the external hydrostatic pressure so that the buckle travels back along the pipe. While this case is usually within the province of the designer, the constructor must make sure that this cannot occur; else the entire line could be lost. Where calculations show this to be problem, buckle arrestors in the form of thicker pipe or reinforced pipe are installed at intervals of 1000 m or so. For example, wrap-around plates may have been pre-installed.

Occasionally, a line may be damaged after it has been successfully laid down. Often this is due to an anchor dragging into the pipeline. It may even be an anchor from your own spread, that is, the barge or boats, or it may be from another contractor working on the same platform. The line must be repaired. One method is to use a hyperbaric chamber (a "habitat") lowered down over the line and centered on the junction or repair point. Compressed gas is used to expel the water; the divers descend to make the weld in the gas atmosphere. The selection of the appropriate gas mixture is critical in order to ensure the proper weld quality. Such a repair in 100 m of water, for example, may require several days. It may be tended by the large barge, but if the sea state permits, a smaller support vessel may be used.

The repair procedure consists of accurately cutting the lines and beveling their ends. A template is made to ensure an exact fit of a "pup" (a short, specially cut pipe section), which is fabricated on board and lowered down for welding. After the welds are completed, the joint is coated for corrosion protection. X-ray is usually not practicable, and reliance must then be placed on visual inspection and magnetic-particle or other NDT techniques to verify the quality of the weld.

Wet-welding techniques have been under development for many years; the problem, of course, is ensuring a weld that will be safe under the working pressures, which typically are 10 MPa or so. At the present time, there is not universal confidence in the quality obtainable by wet-welding techniques, but development continues. Wet-connector devices are now commercially available, permitting some repairs to be made underwater mechanically.

In shallow water, the damaged section can be brought to a surface for a dry weld. The line should be empty, if possible, and long length brought up so as not to exceed curvature limits in the pipe. For this reason, many shallow water lay barges are fitted with davits along one side, enabling lifting from the entire length of the barge. The derrick crane may also pick from the stern while the pipe transfer crane picks from the bow. Where this curvature is still too great, floats or buoys may be attached to give positive buoyancy along appropriate lengths of pipeline.

As the line is brought up, there are, of course, length-compatibility problems in all but very shallow water. It is sometimes necessary to cut the line, thus flooding the pipe. Once brought to deck level, the ends are beveled, a pup fabricated and installed, and the line laid back. Now the new line is longer than required, so it must be laid back to lie in a horizontal curve on the seafloor. The pig is now run to empty the pipe.

Hydrotech has developed a two-piece diagonally flanged coupling which can be sleeved over the two pipe ends and then rotated to accommodate a difference in angle. A three-piece coupling can accommodate up to 15° misalignment. After the joints are bolted up, they are seal-welded, using a dry chamber filled with insert gas.

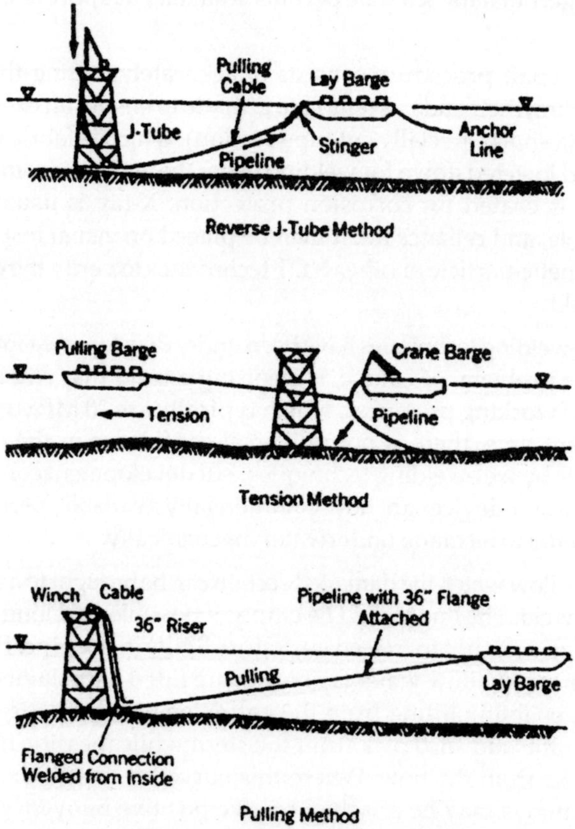

Figure 9.4: Riser Installation Methods

Vickers has developed an explosive-welding method which is especially suitable for tie-ins between pipelines and risers. The explosion is installed inside the pipe, forcing it out against the sleeve to give a solid intermolecular bond. The reliability of this method has not yet been fully accepted.

Advanced "pull-in and latch" connectors have been developed for deep-water connections.

1. In shallow water, the line is packed up by the davits along the starboard side. The derrick picks the riser so that it hangs just off vertical, at the proper angle to the pipe. The welded joint is made. The riser and line are then lowered

back down to the seafloor, the riser coming into position along the jacket to which it is now clamped. In moderately deep water, it may be necessary to add on to the riser from time to time, the so-called "stovepipe" operation. As the riser and pipeline are being lowered, the riser is stopped off from the position and a new length of riser added.

2. For smaller lines, such as flow lines, J-tube risers are built into the platform. The laying is started from the platform; the pipe is pulled off the lay barge and up into the riser tube by a line from the pipe and to a winch on the platform deck. The J-tube bends the pipeline in a permanent but controlled deformation around the 90° bend and then straightens the line with a small reverse bend.
3. For deeper lines, risers are pre-installed on the platform. Alongside is a riser pull-in tube. In some cases, a line may be led out through the riser. This line is then run to the lay barge. As the laying starts, the pipe end is pulled off the platform and to the mating joint, using a winch on the platform to pull in the line. Initial connection is made by bolted flange, followed by internal or external welding as described earlier.

Friction on the Launching Ramp

Friction on the launching ramp can be reduced by the use of rollers or small railcars to support the pipe. The pipe here is in the air, thus having its full weight exerted on the ramp. Movement seaward can be helped by the use of side-boom cats or by an assisting caterpillar-tread tensioner being used in reverse to push the pipe out. As noted earlier, initial sections of pipe may require restraint by use of a holdback winch.

Once underwater, the empty line has only its buoyant weight. This must be slightly negative. This results in friction on the seafloor. It is this friction which the pulling barge must overcome.

Friction coefficients have been measured in the range of 0.4 to 0.5 for the dynamic, moving condition, but rise to 0.6 to 0.8 when the pull is stopped to weld on a new section. Conservative values up to 1.0 are often used in planning since if the line cannot be moved, it will be a total loss.

The pipeline needs enough net weight to be stable on the seafloor and not move laterally. The amount depends on the surf, current, and seafloor conditions, but typical values for coastal lines range from 0.20 to 0.66 kN/m (15 to 50 lb/ft). For bottom-pull installations on the seafloor which do not have to have stability in shallow water, net weight can be significantly reduced. It is the total friction force developed when the line is fully laid that limits the length that can be pulled by this method. If we assume a net weight of 0.3 kN/m, a friction factor of 1.0, and a winch having 1500 kN of pull on the full drum, then the maximum length that can be pulled with a single line is 4500 m (15,000 ft). This can be slightly extended by making short pulls at the end to keep the winch drum half-full, since the winch can apply more force under this condition.

By using two parts of pulling line and a sheave at the nose, the potential overall length of line can be doubled. However, the risk of jamming of the line in the sheave makes this solution acceptable only if the required force cannot reasonably be provided with a single-line.

The bottom-pull method is extremely sensitive to weight and displacement tolerances, since the net weight, a small value, is the difference between two large numbers. Therefore, a great care has to be taken to control and monitor the actual values.

If the single-line capacity of the pulling winch is 1350 kN, this means that if the tolerance shown are actually realized, it may be inadequate. A double line around a sheave will be required. If due to sanding in, the friction rises to a factor of 0.8, then the total force available of about 2700 kN will be just barely enough.

In order to measure and check tolerances, the following procedure has been found effective. Three random but typically appearing pipe sections of 40-ft nominal length are selected for weighing. They are placed in seawater for 24 hours and then lifted out and accurately weighed. Their steel pipe wall thickness is calipered and the diameter measured. The circumference of the coated sections is also measured at three points along the length. Then all subsequent pipe sections are measured for steel pipe wall thickness and diameter and measured with a tape for a circumference of the coated section. The above will enable net weights to be calculated within 2 to 5%, if the pipes are relatively uniform.

Once the pipeline has been pulled, it is folded for stability. A test plug is fitted on the inner end to permit hydrostatic testing. After testing, one pig is then activated by compressed air to empty the line.

The second pig is there for potential problems such as buckling. Buckling of a pulled pipe does not occur in the vertical plane, as it does when laying from a lay barge, but in the horizontal plane, usually due to longshore currents.

When even longer lengths of pipeline are pulled, three options are available:

1. Increase the winch capacity. Conventional winches have upper limits, so a linear cable jack with spooling drum may be used, increasing single-line pulling force to 4500 kN (1,000,000 lb). Wire rope of this capacity working strength would be excessively large and moreover might have excessive bottom friction itself. For a bottom-pull pipe project across Spencer Gulf in Australia, the contractor used high-strength, 10-in, pipe, empty, as a pulling line, and a linear jack, in this case fitted with pipe grips, as the pulling winch.
2. Pull one line out its maximum distance. Pull a second line out beyond it, so that the inner end of the second is at the outer end of the first. Make a connection by flanges or by lifting up the two lines above surface for welding. This procedure was employed for an offshore terminal off Antigua. The ends of the two lines were brought as close together as possible, a "pup" section fabricated from template measurements, and flanged connections made.
3. Decrease net weight and apply very careful monitoring. On a long pull across the Bay of Trieste, Yugoslavia, each length of pipe was weighed in water in a special tank. This enable the net weight on the bottom to be decreased to 0.1 kN/m (7 lb/ft).

The net weight can also be reduced, hence the required pulling force lowered, if floats are attached to the line. Oil drums have often been used in the past but prove rather crude and unreliable. Polyurethane floats can be accurately designed and attached by straps. This increases the risk of problems during installation, since the floats add significantly to the drag force from waves and current. On occasion they have been torn loose; if numerous floats are torn off, the line may become too heavy to move, and thus end up as a catastrophic loss. This occurred on an early line offshore Libya. Assuming satisfactory installation, the floats must later be cut off. While divers were used in the past, mechanical equipment has been designed to travel along the completed pipeline, serving the straps. ROVs, equipped with cutters, have also been used. Despite the potential problems, floats are a viable and accepted solution for heavy pipelines.

The reason a swivel is installed at the nose of a pulled line is to prevent the natural twisting of the wire line under tension from imparting twist to the pipeline. The nose is made buoyant and sometimes shaped like a sled to prevent it from digging in as it is pulled. A pendant and buoy is often fitted to the nose to enable its progress to be observed visually and to facilitate recovery in the event of problems. Often a flanged elbow is incorporated in the nose piece to facilitate later connection to a riser or hose.

A pulled pipeline normally will follow the path of the pulling force, so that it is usually possible to pull around a long radius curve. However, on a hard sand bottom this is not always true, and in such a case the line may drag sideways. One solution is to make the line heavier, that is, increase the net weight to give it more stability. In other cases, lines to anchors have been rigged so that periodically the curve is pulled back to the position. Solutions such as this, or trying to pull around a pile, often results in buckling in the horizontal plane. A better solution may be to spread crushed rock on the seafloor at the zone where the bend mist take place in order to get more lateral stability (*i.e.* higher local friction).

Another solution is to secure several shots of chain inside the pipe, held at the location of the bend or in the heavy surf zone by a line running back to the holdback winch's second drum. This way, the extra weight of the chain stays at the critical location while the line is pulled past. In one unfortunate contract, the owner furnished coated steel line which was almost in equilibrium as to buoyancy, having only a few pounds of negative net weight. The contractor, faced with an oncoming storm, decided to go ahead with the pull and try to get the line in place and flooded before the storm hit. As indicated in previous Chapter, long-period swells run out ahead of storm. Therefore, as the pull was in progress the swells came in at about a 45° angle. While they are refracted around to the normal in shallow water, there was still a net volume of water to be displaced to the south, resulting in a strong wave-induced long shore current. This bowed the pipeline out until it buckled. The contractor aggravated the situation by attaching a line at the buckle leading to a tractor on shore; this attempt to pull sidewise broke the pipe and the entire line had to be abandoned. On the next try, the contractor used an ingenious trick. He filled the line with rock salt to give it weight, pulled the line out properly, then washed the salt out.

For pulling around a curve, saw cuts have been made in the concrete coating to reduce the stiffness.

One more tale of disaster will be given to illustrate the interaction of the hydrodynamic weight, buoyancy, and structural aspects. This line was in the Bay of Fundy, with 10-m tides twice a day. The concrete coating was reinforced only with a very light mesh resembling chicken wire. The launching ramp terminated at high tide; the line was to be pulled across the long tidal flats at high tide and out to an offloading buoy. As the pilling operation commenced, the offshore anchors of the barge slipped. By the time these were reset, the tide was falling, exposing the line on the mud flat. Now as they pulled, they had the increased friction of the pipe's air weight burying the line in the mud flat and developing excessive friction. They then held on until the tide came in and released the pipe from the mud, but now the winds and sea were kicking up and the line bowed laterally. This caused the concrete weight coating to crack, the light wire mesh to break, and the line to float. Eventually the line broke and ended up on the bench, where it had to be abandoned due to multiple kinks and buckles.

Unfortunately, variations on this theme have occurred elsewhere in the world; for example, one of the first submarine lines to Kharg Island, Iran, also reportedly ended up as "spaghetti" on the bench. Recently, a number of cases of "float-up" of pipelines have occurred in the North Sea, apparently due to braking off of large segments of concrete coating due to wave-generated pore pressures within cracks and delaminations and to flexing of the line under vortex-shedding movements.

A gas line pulled across the Strait of Magellan became exposed as it crossed the beach, due to long shore currents and storm waves.

Trenching was impracticable in the cobbles and gravel. The pipe was eventually covered with large riprap. Use of heavy-weight aggregate was another solution that was considered.

Pipelines crossing sandy and/or silty beaches in such widespread locations as Bass Straits, Australia and the landfalls from the North Sea have become uncovered due to the pore pressures generated in the underlying sands combined with the uplift as the top of the pipe became exposed.

The lessons are clear. To all the recommendations regarding the weight and buoyancy control must be added to the need to ensure adequate reinforcement in the concrete weight coating, to backfill over the pipe in areas of strong current, and to special protection in shallow water and the beach zone.

Several clever schemes have been developed to extend the use of the bottom-pull method. One of these involves the pre-laying of a wire line cable on the seafloor, with a series of heavy anchors at the outer end, so that the line's full strength can be developed. Then a special pulling head is attached which incorporates hydraulically-actuated cable grip hoists. One arrangement has these jacks inside the nose piece, feeding the line through a watertight sleeve. This enables the hydraulic lines to be fed through the pipe. In this scheme, the pipeline "swallows" the cable as it pulls itself out along the line. In this system, however, the weight of the pipeline is increased by the weight of the line.

In another arrangement the cable grip hoists are located outside the nose and the pipeline crawls over the cable, pulling itself forward. This system maintains the weight of the pipeline, and the frictional resistance, at a constant value per unit length.

Another system, developed by Dr. William Buehring, also is based on the use of a pre-laid cable. In this scheme, after the cable is laid, a drilling vessel takes in on the bright of the line by direct vertical pull, thus obtaining the mechanical advantage from the geometry as well as the guidance from the pre-laid cable.

Use of one of the above systems of pulling along a pre-laid cable enables the line to be laid in a moderately sinuous pattern, so that it can be laid around outcrops and steep changes in bathymetry.

The bottom-pull method has been successfully extended to the installation of relatively short deep-water installations such as

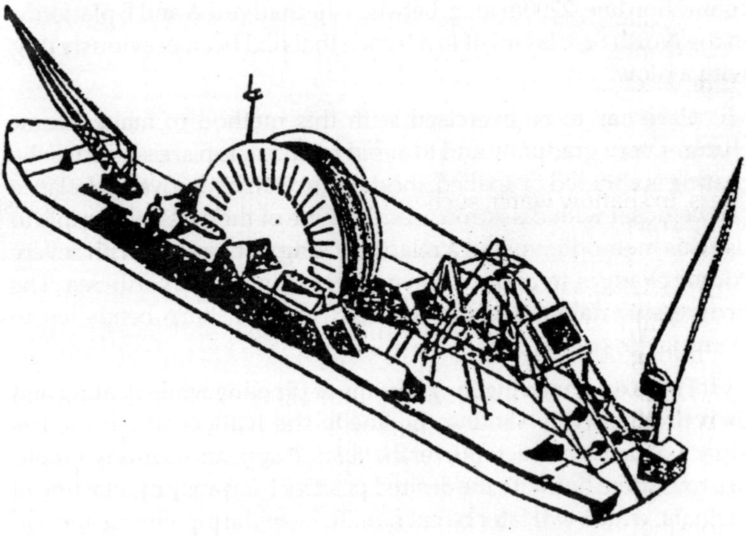

Figure 9.5: Apache, Pipe-laying Reel Ship

interconnecting lines between platform and flow lines. The pulling force is usually that of a large tug and hence is limited to the bollard pull which the tug can exert, with the maximum force being in the range 80 to 150 tons. Being laid in deep water, out of the surf zone, the net weight on the bottom can be reduced to a bare minimum, say 0.2 kN/m.

In 1983, a 4-km (2.4-mile) long bundle consisting of a 12-in. (300-mm) diameter oil line and a 4-in. (100-mm) diameter fuel gas line was pulled off from Ninety-Mile Beach in Bass Straits, Australia, and towed 100 km to connect Fortesque and Halibut platforms. The launch from the beach, which included the jointing of sections, took 21 hours, the bottom tow 33 hours. To reduce friction force and to prevent digging in, 500-m (1600-ft) long sections at each end were buoyed with pontoons to raise above the seabed with slight positive buoyancy. This enabled the new bundle to be pulled over an existing line. As the tow approached the platforms, the end sections were flexed laterally, using winches on the platforms, until mating fittings on the ends of the pipeline were mated with receiving fittings on risers from the pipeline. ROVs were then used to disconnect the 3-in. (75 mm) tow cable and all 88 pontoons.

A similar method was used to install the 36-in. (900-mm) connection line, 2200m long, between the Statfjord A and B platforms in the North Sea, laying it in a trench that had been previously dug with a plow.

Care has to be exercised with this method to make course changes very gradually and to avoid rock outcrop areas, since if the coating is abraded or spalled, the delicate weight- buoyancy balance can be upset with disastrous results. One of the earlier attempts to use this method involved a relatively long bottom tow with severe course changes to avoid known minefields in the North Sea. The progressive damage to the coating from the sharp bends led to eventual loss of the line.

The idea of moving long lengths of pipeline while floating and then progressively sinking the line to the seafloor at the site has attracted many contractors over the years. It appears relatively simple: to provide the line with the desired positive buoyancy by attachment of floats, which will later be cut free. To keep the pipeline in line and prevent buckling due to waves, wind, and so on, one boat tows while a line to shore, a sea anchor or a second boat astern acts as a drag. This keeps the line under tension.

Unfortunately, this method of floating has a number of serious drawbacks, which can also be overcome by through engineering; even then there may be excessive risk. The first problem is that of waves acting on the floating line, causing it to "snakes" in response to the short-crested waves. This may damage the coating and cause it fall off; it turn, weight balance and stability in service are affected. Keeping the line under tension minimizes the dynamic bending. Even with the line in tension, the lateral and vertical forces will alternate over many thousands of cycles and can eventually lead to coating damage. Such a line must obviously be towed out in calm weather. It is very susceptible to even small storms such as squalls. A number of offshore lines have broken up and been lost in this way.

The second problem has to do with the attachment of temporary floats. For moderately long line, there will be hundreds of these. Under the wave action, some attachments may fatigue and fail, thus leading to local areas where the line takes on an increasing sag.

The third and most serious problem to be overcome is that of ballasting down to the seafloor. If floats are cut loose at one end, that end will bend downward sharply and may buckle. The same or worse can happen if attempts are made to introduce water ballast into the line. It will run to one end or a low point; this will cause the rest of the ballast to run to the same spot, and the line will bend and buckle.

Successful installations of relatively short lines in calm water have been made by the flotation method, although most of these were in protected or semi-protected waters. In shallow water, such as river or estuary crossing, the line may be lowered from a series of barges, with multiple control points. Attempts to carry out the surface float method in the open ocean, even in relatively calm seas, have usually failed, an example being an oil import line off Dakar, West Africa.

The essential point is that the entire sequence must be thoroughly engineered to ensure success. Adequate redundancy must be provided to ensure that the loss of one or a few buoys does not lead to progressive failure.

Floats pulled underwater near the bend may collapse progressively. This phenomenon has led to a well-engineered system of air-filled rubber (neoprene) bags, which collapse with depth.

To facilitate connections at the platforms, as the line nears its final location, a line from a J-tube on the platform is affixed to the pipeline end. The buoys at the end are progressively released, while a winch on the platform pulls the end down, either to mate with the J- tube or run on up into the J-tube to deck level.

The controlled underwater floatation method has been developed to overcome some of the deficiencies of the surface flotation method described in next paragraphs. In this method, the pipeline, having slight net negative buoyancy, is towed at a depth of 5 m or so below the surface, where it is much less affected by local waves and not at all by wind.

Support from the line is by hinged or articulated spar buoys attached at frequent intervals. These provide a relative constant upward force, giving a "soft" response, that is, not very responsive to the changes in sea level due to short waves. With their small waterline plane, they do not respond significantly to wind driven waves, and the frequency of response of the system becomes very long (over 1 min). This system therefore virtually eliminates the first major problem of flotation.

The line is kept under tension, with one boat towing, another acting as a stern drag. Upon arrival at the site, the line must be lowered to the seafloor. This is done by cutting off every second spar, by a diver or ROV, while still keeping the line under tension. Of course, when the line is on the seafloor, the remaining spars are also removed.

Continued efforts to develop a reliable method for transport and installation of lines have led to the development of a number of ingenious methods, many of them developed by R. J. Brown. One of these is the "controlled above-bottom pull" method in which the line itself is designed for slight positive buoyancy. Short lengths of chain are attached at frequent intervals to give the overall combination negative buoyancy. Thus it is the end of the chain which drags on the seafloor, not pipe.

The chains automatically control the net underwater weight of the combined system. If the pipeline tends to rise, it lifts more chain off the bottom; if the pipeline tends to sag down, more chain rides on the bottom, reducing the downward pull on the line.

The friction force is now determined by the weight of the "tails" of the chains which drag on the seafloor. The length of tail is in turn determined by the variations in bathymetry over a short distance and the safety required to offset tolerances in weight and buoyancy. While these can be calculated ahead of time, once the pipeline has been launched in relatively shallow water it can be inspected by divers and adjustments in chain length made before pulling out to deep water.

The attachments of the chain to the pipeline must be properly detailed to prevent chafing and abrasion. A week link may be installed to ensure that should a chain snag on an underwater obstruction, it will break free before buckling or crimping the pipe.

The controlled bottom-pull method has proved cost-effective in installations in the North Sea and offshore California, especially for lines of limited length, of the order of 3000 m. During tow, at 3 to 4 knots, the pipeline may "fly" at a depth of 30 to 40 m below the surface. As it approaches location and the tow is slowed, it gradually lowers to just above the seafloor, with the chains dragging the bottom. Then it is pulled into final position and ballasted down onto the seafloor.

A large-diameter pipeline may be employed as a carrier line for several smaller flow lines and cables bundled inside. Obviously, it is essential that the carrier pipe maintain its watertight integrity during the tow. Internal pressurization has been utilized to overcome any leakage. Unfortunately, in at least two recent cases, the combined line has prematurely flooded and sunk, damaging the flow lines being carried inside the carrier pipe. This would indicate that consideration needs to be given to methods of "damage control" of in-leakage due to possible overstress during tow. Appropriate steps might include use of foam, compressed air, multiple pipes, or subdivision.

The controlled subsurface-pull and controlled above-bottom-pull methods have been combined by C.G. Doris for bundled lines to constitute what they have named as the "guide rope method" for placing flow line bundles. The bundle assembly includes both floats and chains. When the pipe is empty, it floats 30 m below the surface; when filled, it floats 10 m above the bottom. Thus, it can "fly" over obstacles, rock outcrops, escarpments, and other pipelines. Filling produces significant bending in the line. This is best controlled by flooding one line in a bundle at a time.

One such bundle of 4- and 2-in, pipes was towed 13 miles just below the surface, and then sunk to the above-bottom mode and the ends pulled into subsea manifold connections, using a guide funnel and pull-in lines through sheaves. Cutting the floats allowed the line to sink to the seafloor.

For installing pipelines between the underwater manifold and the Cormorant A platform in the North Sea, an 8-in oil line plus 2- to 3- in. TFL well test service lines were made up inside insulated sleeves which in turn were placed within two carrier pipes, one 26-in. line and one 24-in. line. The bundle was then towed in two 3.3 km lengths, kept at mid-depth by the chains for a distance of 490 km using one tug of 75-ton maximum bollard pull pulling and another of 35-ton maximum bollard pull restraining, that is, keeping the line in tension. The average pulling forces were 50-ton pull, and 12-ton restraint, and the average speed 5.5 knots. Depth below surface was 30 to 100 m. slowing down and reducing the tension allowed the line to submerge to near bottom. When in position and mated, the carrier pipe was flooded.

The carrier pipes were pressurized with nitrogen gas to 15 atm to prevent water in leakage. A full contingency plan was developed to cope with possible accidents or difficulties, such as the snagging of a chain on an obstruction. The position of the pipes as they approached the field installations was monitored by acoustic transponders and side-scan sonar.

The Burial of Pipelines

The burial of pipelines is often required in order to provide protection to the pipeline against repetitive pounding under wave action, the impact of dropped anchors and snagging by trawl boards and to prevent loss of fishing gear by bottom fishermen. Burial of the pipe also permits the pipe to be designed with less net weight (less coating) which in turn reduces the bending stresses during pipe laying.

The "pounding" referred to above is especially serious in the surf zone, as well as in shallow water where vortex shedding by wave-induced currents can cause alternate rising and lowering of sections of the line, leading to fatigue. Concrete coating can be ruptured and break off, allowing the line to rise. The same phenomenon can occur due to high currents alone, in locations such as Cook Inlet, Alaska. In the inner portion of the surf zone, the danger may be aggravated by direct wave impact and by abrasion from moving sand and gravel.

Burial can be accomplished by laying the line in a predredged trench or by subsequent trenching after the line is laid. A similar protection may be given to a surface-laid by covering it with rock.

Where underwater sand dunes are migrating, as in the southern portion of the English Channel, then predredging by a trailer suction hopper dredge has proved practicable. The line is predredged to a stable elevation, and then the pipeline is laid.

Through the surf zone, a variety of solutions are employed. At benches where the surf and longshore currents are relatively mild, a hydraulic dredge may be employed to overdredge a channel through the beach. The line can then be pulled ashore from the lay barge, allowing the sands to naturally backfill the trench. However, where the beach is subjected to heavy pounding from storm surf, over a period of time, the iterative raising of the pore pressures in the sand may jack the pipe up and out to exposure. Thus the pipe must be sufficiently heavy in this zone so that in service it remains stable. This may require extra jacketing, the use of the double-pipe concept, pipeline anchors, or high-density backfill over filter fabric. Instances of such raising and exposure are reported from such widespread areas as Cook Inlet, Alaska, Ninety-Mile Bench on Bass Strait, Australia, and the Strait of Magellan.

Another method, of course, is the use of a sheet pile cofferdam through the surf zone, which keeps the trench open while the line is pulled through it. Finally, a tunnel or tube can be pre-constructed through the surf zone. This can be concrete or steel pipe pre-laid in

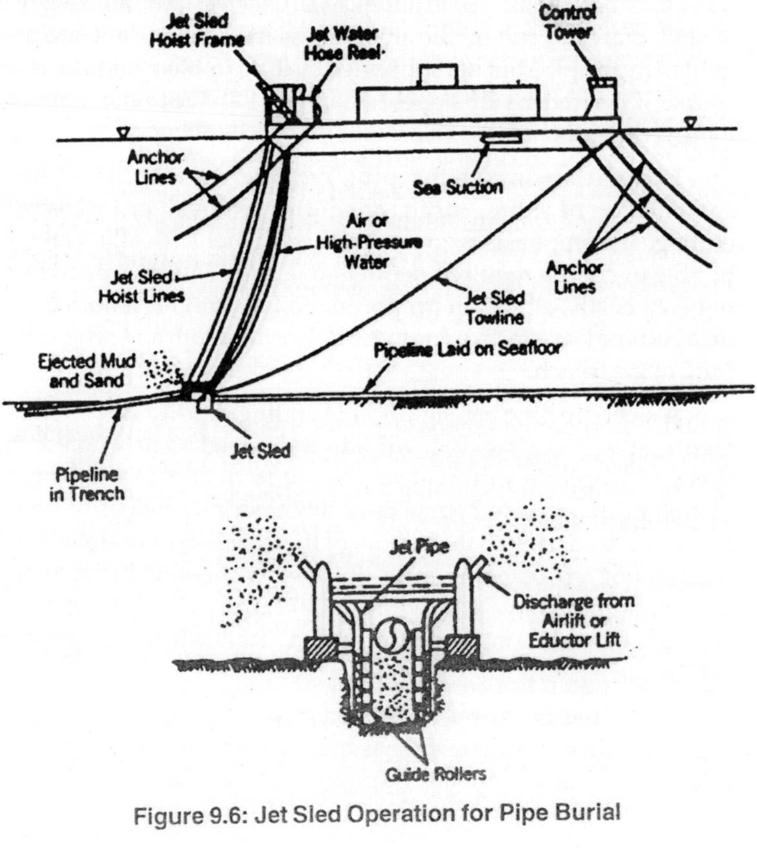

Figure 9.6: Jet Sled Operation for Pipe Burial

a cofferdam, a directionally drilled hole with a casing of large-diameter pipe, or a precast concrete tube jacked in.

A large, precast concrete submerged tunnel or tube was constructed at Cove Point, Maryland, through which an LNG line was run. Similar precast tube segments were installed on the west coast of Norway, on a very exposed rocky coast. The Statpipe Gas Line was then pulled through the tunnel.

For burial of pipelines in deeper water, trenching has most often been carried out by a jet sled, designed to be guided by the pipe and to excavate the soil beneath it so that the line will sink below the seafloor. The jet sled may be designed to run on the pipe, using rubber wheels. The machine must be designed to ensure that its tires cannot damage the coating. In one case in southern California, repeated running of the jet sled over the line broke up the coating to the extent that the line had to be replaced. Other jet sleds are designed to skid, crawl, or run on the adjoining seabed, being centered and guided by the pipe but not supported by it. A problem here arises, of course, if the trench side slopes become too flat due to encountering loose sediments.

Excavation under the pipe may be accomplished by a combination of jetting, airlift, or eductor removal, or mechanical cutting. The emphasis is on powerful equipment. Ideally, it should be able to cut the required depth in one pass. Multiple passes not only are costly, almost in proportion to the number required, but may become less effective due to the increasing depth and progressive infill of the trench.

A steel pipeline has significant bending rigidity and strength. Hence, it will not move down into a dredged trench unless the dredged length is sufficient to cause it to deflect to the bottom. Therefore, the trench when cut must stay open long enough to enable the line to feed itself to the bottom of the trench. Fortunately, this is usually not a problem in most deep-water pipeline installations, since bottom currents and sediment infill are usually limited over a short period of time in the relatively calm seas in which the operation will be carried out.

Power requirements for jet sleds and trenching machines are high. As much as 32,000 HP has been used to power the jets and eductors of a large pipeline burial system used to trench boulder clay in the North Sea. For the pipeline bury barge *Creek*, eight engines drive jet pumps to produce 76 m^3/s at 17 MPa (2500 psi) pressure. Eductors are more efficient than airlifts in the removal of material.

In most cases, once pipelines have been trenched and lowered to their designed elevation, natural sediment transport has been counted on to backfill the trench. If a pipeline is to be backfilled by dumping or placing sand, care has to be taken that the flowing sand, which is temporarily a high-density fluid, does not raise the line out of the trench. This has occurred on both small and large pipelines, to the great embarrassment of all concerned.

Another method of pipe burial involves the principle of liquefaction. By introducing water and air under the pipeline along a length, the sand is "fluidized", allowing the pipeline to sink of its own weight. Obviously, this works best in easily liquefied materials such as fine sands and silts. Vibration applied to the pipe -as, for example, from inside–aids the process. This method has so far been used, to the author's knowledge, only for relatively short lengths of line (*e.g.* through a beach zone).

Ever more sophisticated trenching and burial equipment has been developed, such as mechanical trenching machines. Like the jet sleds, these are guided by the pre-laid pipeline, but they take their support from sleds on tracks at the sides. Rotating trenchers excavate beneath the pipe and throw the material to the side.

The newest form of trenching is that of plowing. A monstrous plow is pulled along the seafloor, stabilized against tipping by widespread outrigger sleds or, more recently, rotating shares (wheels). The plow digs the trench, forcing the excavated material up on the sides.

This development was pioneered by R.J. Brown and has proved very successful in the firms clays of the North Sea. The plows have been designed for the soils expected to be encountered, heavy in North Sea clays, lighter in recent sediments in the Beaufort Sea. In the Bass Strait, Australia, an 80-ton plow was used to post-trench the line, with the plow designed to ride along the pipeline without touching it except at curves. This pipe dug a furrow up to 1.2 m deep in sand and partially cemented sandstone.

Towing forces generally are one to two times the weight of the plow. Traction is provided by a large, dynamically positioned towboat for relatively light plows or by an offshore derrick barge for heavy plows. It might be practicable for the plow to be rigged with a large hydraulic linear jack to pull itself along a pre-laid wire rope or even an uncoated pipeline.

One of the most spectacular uses of the plow to date was on the Northwest Shelf of Australia, where an enlarged version of the plow weighing 380 tons trenched a 1- to 2.3-m-deep trench for 118 km of 46-in. (1150 mm) O.D. line in only one month. The plow had to dig through limestone and caprock, requiring up to 460 tons pull, whereas in softer materials, sand plus silt and clay, 250-ton pulling force was sufficient. Plowing rates reported were 15 to 45 m/min in sand, 10 to 20 m/min in sand over rock, and 5 to 10 m/min in limestone. Difficulties arose principally in soft material, where the plow dug itself in too deeply.

The plow was pulled by a large offshore pipe-laying barge, using a chain-pulling line and developing its reaction force from the barge's anchor lines. The sequence of start-up was as follows. The two mating (positioning) cones were lowered over the pipeline, being spaced apart 40 m by a strut. Divers guided them to seat over the pipe on the seafloor. The plow was now floated to position over the pipe and ballasted down. Removable female sleeves fitted over the pre-placed mating cones. With the plow now accurately in place, the pipe was pulled up into roller guides at each end. The plowshares were then lowered hydraulically to meet underneath the pipe, where they are clamped together. The forward part of the plow rides on the two outrigger crawlers, which are Caterpillar D-9 tractor underbodies.

The derrick barge has hydraulic controls which enable it to raise or lower the plow in relation to the crawler, thus regulating the depth of cut.

The plow concept uses a long beam to automatically even out irregularities on the seafloor. R.j. Brown has suggested that use of computerized controls, reacting to leading sensors, will be developed in the future to handle even rough seafloor profiles. It is believed that a similar system could be employed to scrape the spoil piles back into the trench should early backfill be required.

Plowing appears especially attractive in Arctic soils where lines will have to be trenched deeply (up to 3 m) in order to protect them from the scour of sea ice pressure ridge keels. For the connection line

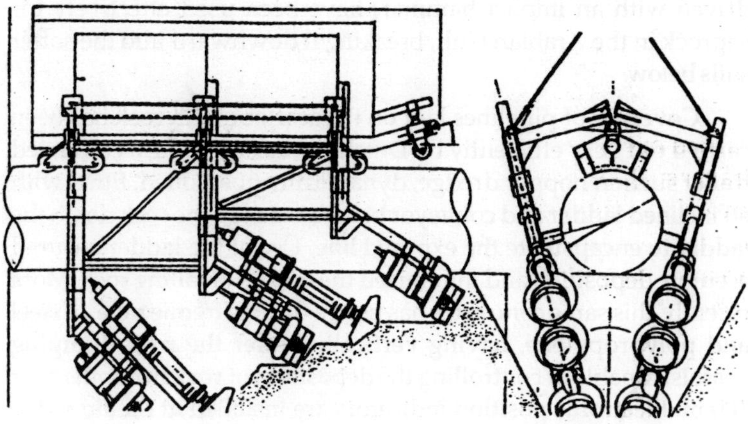

Figure 9.7: Jet Sled Device for Pipeline Burial Uses Multiple Submersible Pumps with Agitators

between the Gullfaks A and Statfjord C platforms, the hard boulder-clay required several passes of the plow. Hydraulically operated grader blades were designed to push the spoil banks aside, leaving a flat level surface from the following passes of the plow. On the Heimdahl pipeline in the North Sea, the 115-km pipeline was post-trenched by plow in only 11.5 days. The plow weighed 145 tons. It was deployed onto the pipeline in 150-m water depth in 19 hours. Recovery after completion required 13 hours.

The most recently developed plows have small shares at the leading end, which ride the previously installed pipe under sensor control, to clear the pipe for the principal plowing action of the larger, deeper plow behind.

When caprock or rock outcroppings must be trenched, it is normal first to break them up with explosives. These can be shared charges laid on the seafloor or drilled in, using high-pressure jet drills or percussion drills. With caprock it is important not to drill through the hard overlying layers, since then the explosion will take place under the caprock, resulting in it being broken into large slabs only, which are extremely difficult to excavate. For overconsolidated silts and in permafrost, high-pressure jet drilling has generally been found more effective than rotary or pneumatic drilling for the placement of explosives.

Rock breaks (huge chisels repeatedly raised and dropped, or driven with an impact hammer) have been used effectively for caprock in the Arabian Gulf, breaking it downward into the softer soils below.

Covering of pipelines laid on the seafloor with rock has been carried out very efficiently by Dutch engineers, using a converted trailer suction hopper dredge, dynamically positioned, fitted with an inclined ladder and conveyor belts, discharge the rock down the ladder to encapsulate the exposed line. Use of the ladder ensured accurate deposition and minimized the impact of falling rock. More recently, this same contractor has used a flexible tremie tube, of steel and polypropylene, having vertically under the rock dumping vessels, capable of controlling the deposition of rock in depths over 100 m. Electronic position indicators are installed at the tip of the tremie.

Sarmac of Italy has developed flexible mattresses of rock-filled fabric which can be placed over submarine pipelines to protect them. They were utilized by Snam-Progetti to cover the gas pipelines from Algeria to Sicily, at depths of 500 to 600 m. They were lowered to the seafloor, using a structural steel frame that was automatically released once the mattress had been placed on the line. These mattresses are also being used to protect pipelines at a location where they are crossed by another line.

Flexible (articulated) mats of concrete blocks were used to cover portions of the gas pipeline off the Northwest Shelf of Australia, where it was laid on a bare rock seafloor.

Pipeline anchors have been used to hold down the pipe at beach crossings and in areas of high bottom currents, especially when the line lies on a hard, bare seafloor. These are usually screw anchors, which are drilled into the soil to secure an inverted U-clamp over the pipe. While these anchors can be installed by divers, this is slow and expensive. Systems have therefore been developed for installing them from a barge directly, using a frame lowered over the pipeline.

For crossing beach areas where high currents and waves have created a deep layer of cobbles, the protection of the pipeline poses significant difficulties. The use of riprap has been previously described. Heavy-density rock such as iron ore is a possible solution, as well as double-jacketing with concrete pipe.

To cross a rocky and exposed coastline on the west coast of Norway, Statoil built the previously mentioned precast, prestressed concrete submerged tunnel.

A protective fence is installed to ensure that rock fall will not endanger divers. Then the site will be leveled by the clamshell. Deposition is permitted on the adjacent lake floor as long as the bucket is lowered to the bottom before opening.

A template frame is then set and pinned to the bottom by drilled-in pipe pile spuds. The template itself is lowered onto the template frame, leveled by preinstalled shims. There will be three transducers on the template. Now the casing and down-the-hole drill are set and entered into the template by ROV sonar and video control and 15-m-deep holes are drilled. The drill cuttings are raised by air-assisted circulation to the surface, then down a tremie pipe for disposal at the seafloor. After a honeycomb has been established over a 5-m-diameter circular cross section, the template is removed, the interstices broken down by a spud chisel, and the rock fragments removed by clamshell.

Then the process of drilling, chiseling, and excavation is repeated until the design depth is reached. A template is run down the hole to ensure that minimum diameter and alignment have been attained. The casing is then floated to the site with end bulkheads, up-ended by ballasting, and seated into the drilled hole. The pipe has a domed bottom at the lower end and valved inlet and vent on the upper temporary closure. Piping for tremie placement of the fine concrete in the annulus is welded on the pipe at two locations, to disperse the concrete. One set of three will fill from the bottom to midway, the other set will fill the top. Concrete is an oversanded mix, with 10 mm coarse aggregate, cement plus fly ash, and an antiwashout admixture.

After a short length has been concreted and gained strength over 2 or 3 days, the casing is ballasted so that it does not float in the grout. Centralizing fins on the casing will ensure minimal annuals width. The annulus is then fully concreted. The tunnel will now be driven to the intersection, with its invert just below the tip of the shaft casing. Grout holes will be drilled to encase the connection zone and grouted. Then the connecting elbow will be hand-mined.

Chapter 10

The Subsea Supplies and Reliability

Introduction

The subsea supply is an important topic which takes a major role in the offshore electrical engineering. To those new to the subject of offshore electrical engineering, using electrical power on equipment which is totally immersed in highly conducting salt water may seem somewhat foolhardy. In actual fact, electrical power has been used in submersible pumps and for underwater are welding since around the turn of century. Providing certain precautions are taken, using electricity under the sea can be comparatively safe.

Permanent arrangements will be required for power to operate valves, switches and sensors on underwater wellhead manifolds and the new generation of subsea production wells which are serviced periodically by special vessels. Where power and control are required at the manifold from a platform some distance away, an unearthed supply will be needed at a suitable voltage to compensate for the impedance of the subsea cable. It is usual, however, for power for operation of valves etc. to be transmitted hydraulically with electrical actuation from a secure (UPS) AC supply on the adjacent platform.

It is recommended that all AC supplies used are three-phase and are isolated via the delta connected, unearthed secondary windings of transformers. This provides the best scope for incorporating protection devices, and single earth faults can be arranged to operate alarms whilst the circuit remains in operation.

Now let us try to understand about the needs of diver. The main requirement for the North Sea diver is for heat, both for body and for respiratory gas heaters.

The design of respiratory gas heating, if fitted to the diver's suit, must be such that there is no possibility of electrocution through the mouth. If the diver's suit is heated electrically then the elements of heated undergarments should be screened such that there is no possibility of the heating elements touching the diver's skin, or the elements shorting internally and causing localized heating.

Both respiratory gas heaters and electric body heaters should use a low-voltage DC supply no greater than 24 V DC. Heater elements, including supply cables, should be screened, with the screen continuity continuously monitored.

A line insulation monitor (LIM), designed on failsafe principles, should be incorporated into the supply circuit to detect any deterioration in the supply cable to heater element insulation resistance. The sensitivity of the LIM should be such as to ensure a tripping response in a few milliseconds to a significant fall in resistance.

By far the most common form of diver's suit heating in the North Sea, however, is the hot water machine from which heated sea water is pumped from the surface. The heated water is directly in contact with the diver's skin and escapes from wrist and ankle bands in the suit. There is a remote possibility that an earth leakage current could travel down the heated water via the hose and be earthed via the diver. However, if the heater element tank in the hot water machine is adequately earthed, there should be no danger, especially as the incoming cold water also provides a conducting path.

In offshore electrical engineering, the reliability analysis takes an important role. It is very important when carrying out any form of reliability analysis that specific goals are selected. The availability of an electrical supply at a particular point in a system will be very much dependant on the configuration of the system and where that point is located in it. Therefore, a vague instruction such as 'qualify the reliability of the supply system' is very unhelpful, and it would be necessary to discuss with the client concerned what particular supplies were of particular concern and then to consider each in turn. For example, if the object was to maintain full production then only the equipment required to maintain supplies to the production switchboards would need to be considered. However, if supplies to essential equipment only were being analysed, the availability of all generation and all routes which could be used to connect it to the essential equipment would need to be taken into account and, hopefully, in comparison a much higher figure for the availability of supply would be obtained.

The Diving Chambers and Inductive Couplers

As the lives of divers may depend on the electrical supply being available at the chamber, particularly if deep diving involving compression chambers is being undertaken, the electrical supply systems of the associated support vessels must be of the highest integrity and reliability.

The equipment used inside diving chambers is, where possible, to be designed using the principles of hazardous area intrinsic safety. This is for the following reasons:

It will prevent ignition should be flammable well gases be present.

It will help to prevent fires in a chamber with an oxygen enriched atmosphere.

It will ensure earth fault currents are limited to very low values.

Table 10.1 shows the effects on the human body of various magnitudes of AC current. From this it can be seen that values beyond 18 mA are dangerous. In salt water the resistance tends to be low, and very low voltages are required to achieve this threshold value. All electrical circuits within the chamber will need to be protected against earth faults by using current operated earth leakage circuit breakers set to trip at not more than 15 mA in 30 ms.

The power supply within the chamber should be low voltage and either direct current (*e.g.* 24 V DC) or possibly at a frequency greater than 20 kHz, to which, apparently, the human body responds in a similar way to direct current. Any sockets used should be the underwater type, which can be connected while immersed in sea water without hazard. Unused sockets should be provided with insulating caps. If no suitable intrinsically safe equipment can be obtained, equipment incorporating other types of protection may be used. Any electric motors in the chamber should be of the flameproof or increased safety type, and designed to withstand stalling currents for long periods without detriment.

Table10.1

Current at 50 Hz to 60 Hz rms milliamperes	*Duration of Shock*	*Physiological Effects on Humans*
0-1	Not critical	Range up to the threshold of perception, electric current not felt.
1-18	Not critical	Independent release of hands from objects gripped no longer possible. Possibly powerful and sometimes painful effects on the muscles of fingers and arms.
18-30	Minutes	Cramp-like contraction of arms. Difficulty in breathing. Rise in Blood pressure. Limit of tolerability.
30-50	Seconds to minutes	Heart irregularities. Rise in blood pressure. Powerful cramp effect. Unconsciousness, ventricular fibrillation if long shock at upper limits of range.
	Less than cardiac cycle	No ventricular fibrillation, heavy shock.
50 to a few hundred	Above one cardiac cycle	Ventricular fibrillation. Beginning of electrocution in relation to heart phase not important. (Disturbance of stimulus conducting system?) Unconsciousness. Current marks.
	Less than cardiac cycle	Ventricular fibrillation. Beginning of electrocution in relation to heart phase important. Initiation of fibrillation only in the sensitive phase. (Direct stimulatory effects on heart muscles?) Unconsciousness. Current marks.
Above a few hundred	Over one cardiac cycle	Reversible cardiac arrest. Range of electrical defibrillation. Unconsciousness. Current marks. Burns.

It is more than likely that surface Ex'd' certification will prove invalid for equipment to be used in underwater chambers, because of the higher breathing atmosphere pressures required for the long-term compression diving. The equipment enclosure will require pressure testing to 1.5 times the normal peak explosion pressure times the working pressure of the chamber (bar A). Intrinsically safe equipment safety margins are also reduced by increase in pressure, and such systems will also require reassessment when used in this environment. Other types of certified equipment may be affected by the higher humidity or by water ingress because of the higher pressures likely to be experienced by underwater equipment.

Now let us try to understand about inductive couplers. If the core laminations of a transformer are cut in half, the primary coil would on one half and the secondary wound on the other, then when the two halves are mated together a rather inefficient transformer is formed. This principle is used in thousands of inductive couplers throughout the North Sea and elsewhere.

The core and the winding are encapsulated so that the sea water at North Sea sea bed pressure cannot penetrate into the live parts of the device. Only the surface of the core, which is ground flat, is exposed. Because of the extra losses produced by the position of the windings and the air gap in the core, more heat is generated than in a conventional transformer, and this must be removed by a core heat sink connected to the stainless steel housing.

The Umbilicals

An umbilical is set hydraulic hoses and signal, control and power cables bundled together, armoured and sheathed to form a single entity and used to provide services from a surface installation or vessel to equipment below the surface, often on the sea bed. Individual cables within the umbilical may also be screened or armoured or both, and the whole umbilical may be strengthened against tension damage during development by the provision of an embedded steel tension cable.

A subsea cable is usually of simpler construction, containing the three main power cores operating at usually between 11 and 33 kV, with telecommunications and protection pilot wires often located in the centre of the cross-section. The cable will be armoured and sheathed overall and may also contain a tension wire. Cables are brought on to a platform via a J-tube, which is a length of steel guide pipe fixed at various points to the installation jacket or hull and curling out at the bottom to receive the cable from the sea bed.

Whether a subsea link is required between two or more platform electrical power systems, or between a surface installation and sea bed equipment, the need for both a very high level of manufacturing quality control and strict adherence to the planned delivery dates is vital. Several very expansive mistakes have occurred in recent years, where umbilicals of 10 kilometres or more have had to be scrapped because of the discovery after completion that vital cables within the umbilicals were faulty. Apart from the financial penalties involved, the time delay in reconstructing the umbilicals threatened to delay their laying beyond the North Sea summer weather window.

The thorough quality control procedures adopted during construction should include the following:

1. Discussion should be held with the manufactures in order to arrive at the optimum cable construction for the particular application. This will include:
   (*a*) Length of umbilical;
   (*b*) Depth of water and hence pressure;
   (*c*) Type of cable laying equipment used, handling conditions and tension likely to be applied;
   (*d*) Electrical parameters required to ensure satisfactory operation of the equipment at either end, which as a minimum should include voltage, current, resistance, reactance, capacitance and attenuation over an agreed frequency range;
   (*e*) Storage conditions.
2. Analysis should be made of samples of all materials used for the minimum performance required. Insulation thicknesses should be more generous than those for cables are used on land. If screen drain wires are required, ensure that these do not have a cheesewire effect on insulation when the umbilical is in tension.
3. Close control of conductor and insulation diameter and thickness of each cable should be maintained during manufacture.

4. Samples of each cable should be taken at the start and finish of each stage of the cable manufacturing run for analysis, and the results of analysis should be known before continuing with the next stage.
5. The final testing of each cable should include insulation resistance, high voltage testing and measurement of total cable resistance, reactance, capacitance and attenuation over a given frequency range. If DC switching or digital signals are being transmitted, low attenuation at megahertz frequencies may be important in order to avoid degradation of the signal over the length of the cable. Such readings may also detect faults, if attenuation readings for known healthy cables are available for comparison.
6. A sample of the completed cable should undergo high-voltage testing in water at the pressure equivalent to at least the maximum depth of water likely in the installed location. The test is carried out using a pressure vessel with a cable gland at each end. The cable sample is passed between two glands, which are then sealed. The vessel is filled and pressurized and the high voltage can then be applied at the cable ends protruding from the glands.
7. Another high voltage test often conducted on a sample is to submerge the sample in a tray of salt water with one exposed conductor in air at either end. A high voltage is then applied between a submerged conductor (or the tray) and the exposed conductor in air. This and the previous test are carried out the maximum voltage available or until a voltage breakdown occurs in the sample.
8. Once all the cables have been successfully completed their testing, they may be incorporated into the umbilical. However, if there is a delay between cable and umbilical manufacture, or the cable has to be transported to a different site, further insulation testing will be required immediately prior to umbilical manufacture to ensure that transits or coiling damage has occurred. It should also be remembered that faults due to poor curing of the (plastic) insulation may appear several weeks after manufacture.

Impressed Current System and Galvanic Anode Systems

In the impressed current system an external source of power, usually a transformer rectifier unit, provides the driving voltage between the anode and the structure to be protected. The negative terminal of the DC source is electrically connected to the structure and the positive terminal similarly connected to the anode. The output voltage of the DC source is adjustable so that the protection current may be varied either by a trimming potentiometer or automatically through a control signal loop to provide the required voltage at the reference half-cell. Transformer-rectifier for this use are usually rated in the range of 20-500 A and from 30 to 120 V.

Impressed current anodes, unlike galvanic ones, must be well insulated from the protected structure if they are to be mounted on it. Where it is found necessary to place impressed current anodes very close to the protected structure, dielectric shielding between anode and structure must be provided.

As it has a uniformly low electrical resistivity, sea water is an excellent medium for the application of the cathodic protection and facilities an even current distribution over the surface to be protected. Bare steel submerged in sea water is easily polarized if an adequate current density is maintained.

Impressed current anodes may be of silicon cast iron, lead alloy, or graphite. Lead silver alloy or platinized anode, although more expensive at the installation stage, may be used to provide a longer life and hence maintenance cost saving. Impressed current anode systems should be designed for a life of at least 10 years. High silicon cast iron corrodes relatively slowly and in sea water, where chlorine is produced at the anode surface, chromium may be added to further improve longevity.

The resistance of a single anode in free flowing sea water is less than that of the same anode in mud or silt, and it may be necessary to install groups of anodes, sometimes mounted on a wood or concrete framework, in order to reduce silting up maintain effective contact with the sea water. Care must be taken to avoid incidental rubbing or scraping of the active anode surface by suspension or stabilizing rods or cables, since the effect of this is to accelerate dissolution at that point. For obvious reasons, suspending the anode by the conducting cable is not advised. The number of anodes is determined by the required anode life, the allowable current density, and the circuit resistance of the system. Anode life will be affected by the operating current density and the total current magnitude. In free flowing sea water, a maximum output of about 10 A m^{-2} is normal practice, giving an anode consumption rate in the region of 0.4 kg per amp per year.

Platinized anodes are generally constructed from a solid or copper-cored lead alloy rod about 12 mm in diameter and 1500 mm long. A layer of platinum approximately 0.005 mm thick is deposited on this. A similar thickness of platinum may be used on anodes of other shapes and other materials, such as platinized titanium, platinized niobium or lead silver alloy. Graphite is not generally suitable for use offshore.

For reasons of economy, platinized titanium anodes should not be used where silt or mud may accumulate, above the lowest tide level, or where fluctuating voltages, single (*i.e.* low frequency) AC ripple or anode voltages higher than 7 volts are present, since the expected long life will not materialize in these conditions. The system design should be such that these anodes are not expected to carry current densities of more than 700 A m^{-2}. The 0.005 mm thick platinizing dissolves at the rate of about 10 milligrams per ampere year, and although the abrasive action of water-bourne sand may reduce this to some degree, a life expectancy of 15 years may be realistically hoped for.

The anodes are usually mounted on supports cantilevered out, but electrically insulated from the protected structure. In shallow waters, seabed non-metallic frames may be used to support the anodes above the mid or silt level. To protect any offshore platform, distributed multiple anode arrangements will be required to be mounted over the members of the complex subsea structure to provide a uniform current density. The size and design of the cathodic protection scheme will depend on the current density required to bring the structure up to the level of protection required. Sea water velocity and oxygen content will affect chemical activity and hence the current density. Statistical data will be required from the platform site, if an accurate figure for current density is to be determined. Following data gives information about the required magnitudes.

Location	Environment	Required Current ($mA\ m^{-2}$)
Persian Gulf	Sea water	20
	Below mud level	50
North Sea	Sea water	50
	Below mud level	90

Now let us study about galvanic anode system. In the galvanic or sacrificial anode system, the driving voltage between the structure to be protected and the anodes is developed directly by the electrolytic potential between the two (different) metals involved. The anodes are usually of aluminium alloy or magnesium, types which are less affected by insulation from oil wetting such as 'Galvalum III' being often preferred. Galvanic anodes may be placed very close to, or in contact with the structure to be protected. Galvanic anode systems should be designed for a life of 25 years.

'Galvanum III' or similar anodes are suitable for use on both subsea pipelines and offshore structures and are particularly suited for use in saline muds or silted-over conditions. On subsea pipelines it is used in the form of 'bracelets'. On subsea structures, bar type anodes are mounted at suitable intervals.

Although used extensively on ships, zinc and magnesium anodes are not suitable for offshore use mainly due to their faster rate of consumption and hence higher costs involved in replacement. Supplementary magnesium anodes may be used in order to provide a temporary boost in polarizing current, however. As cathodic protection starts to operate, a layer of alkaline material is formed on the protected structure by cathodic electrode reaction. Provided this material is not dislodged, its presence reduces the current density required to maintain protection.

On offshore structures the polarization-boost anode may take the form of a 10 × 20 mm ribbon of magnesium attached to the structure and supplementing the permanent system.

Failure Mode, Effects and Criticality Analysis

FMECA is a procedure, which is very useful in identifying all modes of failure and analysing their effects. This procedure, like hazard and operability analysis (hazop) which is beyond the scope of this book, is a method of identifying all modes of failure, analysing their effects and, where possible, evaluating the frequency of their occurrence. The basic FMECA process is the completion of a table with column heading similar to Figure 10.1.

However, before the columns can be completed, it is necessary to draw a block schematic diagram of the system to be analysed at the level of detail required. This level should be low at the initial stage; as it becomes more obvious which components or subsystem are critical, particular areas can be broken down into more detail. At each level a more detailed block diagram should be drawn.

If we take a circuit breaker as an example item on an FMECA, this would appear as a block in the diagram shown in Figure 10.2. A circuit breaker is a particularly difficult component to describe in an FMECA, as it has several functions and several failure modes.

The circuit breaker functions are as follows:

It connects and isolates an electrical circuit as required.

In the event of a fault, it interrupts the flow of fault current automatically in conjunction with a sensing device.

Note that the adequacy of load and fault rating is assumed, as the purpose of this analysis is not to question the system design calculations but to consider the effects on the system of a particular mode of failure.

The circuit breaker has two modes of failure associated with the first function, namely that it:

(*a*) Fails to open when isolation/disconnection is required;

(*b*) Fails to close when power supply is required.

For the fault clearing function there is one failure mode:

(*c*) Fails to interrupt fault current.

If required this could be split into:

(*c*) (*i*) Fails to interrupt make fault;

(*c*) (*ii*) Fails to interrupt break fault.

Lastly, there are faults associated with the breakdown of insulation within the circuit breaker cubicle:

System: Generation Subsystem: Main HV generation Drawing no:			Failure mode and effect analysis worksheet Job no: Job title: Reliability study Client				Page ___ of 8 Issue A Prepared by: ___ Date: ___ Checked by: ___ Date: ___		
(1) Item code	(2) Component	(3) Function	(4) Failure mode	Failure effect		(7) Means of detection	(8) Severity of effect	(9) Compensating provisions or measures in operation	(10) Remedial measures in design
				(5) Local level	(6) Plant level				
1	Generator UCP	a) Trip/ Inhibit Start under fault or pre-set conditions	Unit starts under fault conditions	Damage to unit	Down time due to original fault extended Reduced production	Operator	II	ST/BY unit GT	Instrumentation and controls, plus maintenance procedures and periods deem adequate from information available
			Unit fails to trip under fault conditions	Damage to unit	Initially no effect. Severe unit damage results in loss of electrical output to distribution reduced production	Operator	II	ST/BY GT Unit if available Load shed	—— " ——
			Unit trips under healthy conditions	Loss of output	Loss of output to distribution load shed Initiate	Operator	I	—— " ——	—— " ——
		b) Auto sequence control brings set on load through pre-set conditions	Wrong Sequence	Damage to unit	—— " ——	Operator	II	Second GT unit	—— " ——
				Fire possible	Fire hazard	Operator/ Control system	III	—— " ——	—— " ——

Figure 10.1: Sheet for Use with Failure Mode, Effects and Criticality Analysis

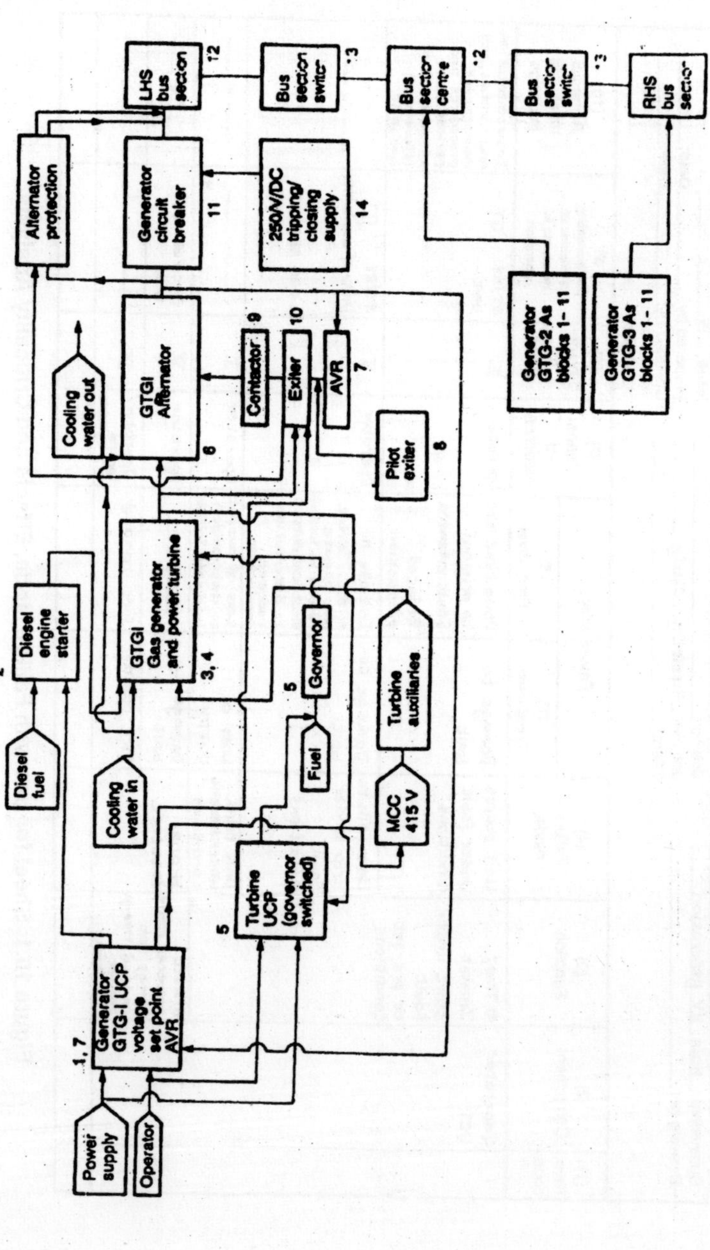

Figure 10.2: Typical Functional Block Diagram for Use with FMECAs

(*d*) Interphase fault in circuit breaker cubicle;

(e) Earth fault in circuit breaker cubicle.

Some of the faults listed may be due to the failure of the trip/ close battery supply or a protection relay. Depending on the level of detail required, however, these components can be included as part of the circuit breaker blocks, or if necessary a new block, diagram may be drawn showing individual blocks for each component in the circuit breaking system.

The failure mode have been identified, the next stage is to describe:

(*a*) The effect on the system;

(*b*) The effect on adjacent equipment;

(*c*) The effect on the offshore installation.

If the system is well designed, there should be no effects on the installation and only one or two effects on adjacent equipment, usually associated with fire. All effects listed should have some compensating provision or remedial measure available to compensate, and this should be listed as shown in figure 10.2 column 9. Where no such provision or remedy exists, column 10 should be used to propose a solution which prevents such an effect or alleviates the problem. At the end of the study all such proposals should be listed.

If a significant number of proposals appear in column 10, it will probably be necessary to rank the associated effects in order of criticality. This is usually done by drawing a *criticality matrix* as shown in Figure 10.3 (extracted from MIL-STD-1629A). This is a graph with axes of severity of effect against the probability of its occurrence. The severity is given four arbitrary categories as follows:

Category I: catastrophic–A failure which may endanger the whole installation and involves loss of lives.

Category II: critical–A failure which leads to prolonged loss of production and may cause severe injury and major equipment and/ or system damage.

Category III: marginal–A failure which may cause minor inquiry or minor system/equipment damage and will result in delays to or partial loss of production.

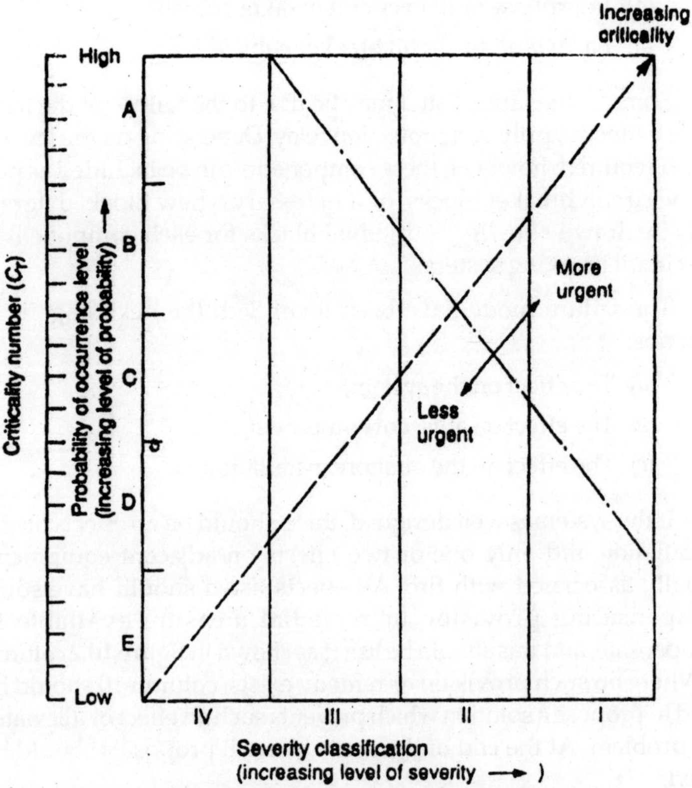

Figure 10.3: Typical Criticality Matrix

Category IV: minor–A failure which is not serious enough to cause injury or any equipment/system damage but which will result in an unscheduled outage for maintenance or repair.

The frequency of occurrence may either be a numerical value obtained from a reliability databank, or a qualitative level assigned as follows:

Level A: frequent–A high probability of occurrence during the time the system is running.

Level B: reasonably probable–A moderate probability of occurrence during the time the system is running.

Level C: occasional–An occasional probability of occurrence during the time the system is running.

Level D: remote–An unlike probability of occurrence during the time the system is running.

Level E: extremely unlikely–An essentially zero probability of occurrence during the time the system is running.

A diagonal line is drawn on the matrix, and each failure mode effect which has no existing compensating provision should be represented by a cross drawn on the matrix. Failure effects appearing in the upper right hand area of the graph have the greatest criticality and hence the most urgent need for corrective action.

Chapter 11

Harbor and River Structures

Introduction

Number of harbor and river structures can be seen in ancient civilizations. Number of bridges has been constructed by Persians and Egyptians. The Phoenicians built breakwaters and crude quays using timber pile cribs filled with stone. The irrigation practices of the Sumerians and Egyptians required river control works. The Greeks had their port at Piraeus. The Romans built an artificial harbor at Ostia, for which they may have used pozzolanic cement. We know for certain that they did build timber cribs and cofferdams for bridge piers as well as pile-supported trestles. Later, the British developed portland cement, and the French conceived the idea of the reinforced concrete. In harbor structures the dominant structural type is the quay or marginal wharf for the unloading and loading of containers. Finger piers, extending normal to the shore, are occasionally used for transfer of petroleum products and as trestles to loading platforms and wharves.

Now let us try to learn a brief information about steel piles. These are either tubular pipes of 400 to 600 mm diameter, or H-piles. The pipe piles present less surface area for corrosion and hence are more readily protected. Although in many soils steel H piles penetrate more readily, in stiff clay the soil plugs between the flanges and the pile act as a partially end-bearing pile of square cross section. The steel pipe pile also plugs, like its larger diameter counterpart offshore. The plugging occurs when the internal skin friction exceeds the ultimate bearing value of the soil at the current tip.

Since the maximum water depths for modern container and cargo ships are about 16 m, and the design pile capacity is only of the order of 200 tons, in most cases piles will range from 30 to 40 m in length. Petroleum terminals require greater depth at the face of the wharf, usually 23 m and hence larger and more heavily loaded piles. Steel piles can be picked (pitched) readily, and placed into the leads. At decks level, the pile may be centered by chocks or, better, a mechanical gate, while at the top a hydraulically operated arm may swing around the pile to center and align it in the leads.

Today concrete piles are almost always prestressed by the pretensioning process. The sizes up to 500 mm are typically square in cross section, those of 600 mm size are octagonal, and from 900 to 1700 mm diameter are round with a hollow core. The precompression in the pile, after losses, is 6 to 9 MPa. Experience indicates that values less than 6 MPa do not perform as well as those with the higher values.

In pitching (up-ending) concrete piles, because of their heavier weight and lower bending capacity, multiple pickup points are required. Their location in the fabrication plant is usually determined on the basis of vertical lifts on a horizontal pile. In the case of pitching into the leads, *i.e.* up-ending, the lines (or slings) will lead at angles which will vary as the pile changes from horizontal to vertical. The vertical components of the various lines will change with the leads of the lifting lines and the pile's inclination. Usually (but not always), the critical case is the initial lift from the horizontal. The degree of prestressing must be adequate for both transport and pitching.

Installation of Piles

After pitching, the steel or concrete pile is then allowed to run down under its own weight. The hammer and driving head are now lowered onto the pile with the pile line slacked and the pile penetrates farther under the weight of hammer. Driving now commences. The first blow may cause the pile to run a meter or more; hence the hammer line must be free in order to keep the hammer and driving head on the pile. Although vibratory hammers may be used to obtain initial penetration of steel piles in cohesionless soils, final seating is normally by impact hammer. Concrete piles are generally driven by impact hammer alone. Three types of impact hammers are in common use today, the single-acting steam hammer (which can also be activated by compressed air), the hydraulic hammer, and the diesel hammer. All three types depend on the mass of the ram impacting on the driving head which in turn transmits the impulse to the pile. In the case of the single-acting steam hammer, the propelling force is gravity alone; in the case of the hydraulic hammer, the gravitational force may be amplified by hydraulic pressure. With the diesel hammer, the force is applied sequentially; first the compression of the air by the falling ram "preloads" the pile, then the ram delivers its impact, and finally the trust of the exploding fuel reacts downward while at the same time raising the ram for the next stroke. As a result, the hammer energies are computed differently; a rough rule of thumb equates the rated energy of a steam or hydraulic hammer to 1.5 times the same rated energy of a diesel hammer insofar as the effectiveness in driving. This is not only due to the differing modes of impact but also to the proportionally lesser weight of the ram in the case of the diesel hammer, which is partially compensated by a higher velocity.

The driving heads are designed to transmit the impact to the pile head. Hence, in the case of steel piles, the head is configured to match the pile section. The below is transmitted steel to steel; no cushion material is interposed. The driving head serves to combine the pile head and prevent it from bulging.

In the case of the concrete pile, a cushion is required to attenuate the blow and extend its duration. The driving head is configured to contain the pile head cushion. The pile head cushion is best made as a 200 to 400-mm thick laminated softwood blocks. Plywood layers may be affixed top and bottom and inserted in the middle to help hold the block together during driving. Experience, supported by dynamic measurements, shows that such a cushion will usually be adequate.

The hammer blow creates a compressive wave which travels down the pile at the speed of sound in the pile material. When the wave reaches the tip, it either causes the pile to penetrate, thus producing a tensile wave in the pile or, if the pile tip is on hard material such as rock, causes a rebound compressive stress wave of twice the intensity. It is this stress concentration which causes the tip of steel piles to buckle, tear, or accordion. The high rebound tensile waves, alternating with the input compression, may lead to low-cycle, high-amplitude fatigue of any welded splices in the steel piles.

In the case of concrete piles, two special stress patterns must be considered in order to permit driving to the required penetration without damage. Unlike steel piles, which are most damaged by the high compression stresses in hard driving, concrete piles are subject to damage in soft driving, such as the initial blows, which produces tensile stresses. Without the proper cushion block and hammer control, horizontal cracking may occur through the entire body of the pile, showing itself initially as a puff of dust at each hammer blow. Continued driving will lead to damage to the concrete at the crack and stresses in the steel beyond yield; prestressing steel will eventually fracture. This phenomenon is also noted when, after driving through component soils which seize the pile in friction, the tip protrudes into weak material or a void below. Such a void may have been created by excessive jetting below the tip. This phenomenon is amplified underwater, where water is progressively sucked into the crack, then subject to high-impact pressure under succeeding blows.

The second special stress pattern is due to Poisson's effects: the lateral bursting of either the head or the tip under axial compression in hard driving. Fortunately, concrete that is well-confined can resists very high compressive forces. Hence, closely spaced spirals at head and tip will prevent damage. Since the problem is greatest at the head, extra confinement of the periphery is needed.

Although, in most cases, pilings for wharves are initially driven through weak or loose sediments to strata of increasing strength and stiffness, other situations are encountered. One of these is where a rock dike has been constructed beforehand, or where a rock hill has been placed on an underwater slope. Experience shows that a compact pile such as a precast concrete pile can successfully penetrate the rock, largely by displacing it sideways. Some deviation can be expected in inclination, *e.g.* 1%, and in position, *e.g.* 300 mm, but a satisfactory piled foundation can still be achieved.

The above has been demonstrated on a number of projects with rock sizes up to 500 mm and dikes up to 10 mm thick. If the underlying sands consist of dense sands or silts for which jetting is required, a jet pipe can be cast into center of the pile and operated continuously during the driving, to prevent plugging.

Piles being driven onto or through rock should have square or blunt ends. Tapered and pointed tips deflect and break the pile. Piles should initially be seated vertically or on the designed inclination. This may require two points of lateral support *i.e.* the drilling head at the top and a template near or below the waterline. After initial penetration and before the full impact of the hammer is delivered, the lower support should be freed. Otherwise, with a third point of support in the dike, the pile may be subject to bending and breakage. As the pile then penetrates, the hammer should be gradually repositioned to "follow the pile".

Now let us learn few things about batter piles. Batter piles are typically used to resist lateral loads. They always act in pairs, one taking compression, the other tension. The present trend in seismic regions, therefore, is to use all vertical piles and take the lateral forces through elastic or elastoplastic bending of the piles in the usual case of a sloping bottom, the shorter piles, being stiffer, will take almost all the lateral loads. Thus they may be made larger or be spaced more closely than those in deeper water.

In some seismic retrofit projects where existing batter piles must be used, ductile fuses have been inserted at the connection; they are based on a connection which deforms in bending under high load and thus limits in compression in the pile which is most highly loaded. This concept has also been used in new construction for quays in Long Beach Harbor, California.

Piles in finger piers, extending out from the land, typically driven with floating equipment, although in areas of extreme tides such as Southern England, piles may be driven by rigs moving out over the top of the completed construction. In quay construction, although the outboard piles are driven afloat, the inboard piles may be driven with a land-based crane, rigged as a pile driver, with a hydraulic operated spotted to position the pile vertically or on a batter.

Due to the close spacing of piles in relation to the width of the typical pile driving barge, the pattern for installation has to be laid out carefully, fanning out from the shore in order to be sure to have access to every pile. This problem is compounded when there are better piles, since the barge must be offset from the designed head location. Some cluster of batter piles, radiating in different directions, may prevent direct driving, so a template must be used, supported by already driven piles and additional spud piles, in order that the new pile may be properly placed.

Batter piles are typically supported in the leads and at the deck level. When they extend down through the water, they are cantilevered and droop until they are embedded in the soil. Thus in deep water they need underwater support, such as telescoping leads. Among the other options available to the contractor are to increase the size of the pile, the concrete strength, and especially the level of prestress. Thus on Pier 23 in san Francisco, 50-m-long batter pile in 15 m of water depth were increased, with approval by the designer, from 450 to 500 mm and were prestressed to 10 MPa, in order to facilitate setting on a 3 in 12 batter.

The first rule is to have a compact section pile at the tip, in order to prevent local damage and progressive buckling and damage. With steel pipe piles, the tip should be closed by a heavy plate. The wall thickness for the lower two diameters should be increased. The tip of the pile can be filled with concrete for two to four diameters. Points and fabricated tips cause the pile tip to deflect, thus producing high bending stresses in pile, so they should not be used.

Alternatively, a cast steel tip protector of the "external" type can be fitted. Such a tip will give a constant inside diameter at the tip of the pipe pile and thus not interfere with subsequent drilling. Then when an obstruction is encountered, a drill may be used to drill through and ahead. Churn drills and down-the-hole drill are the most frequently employed.

Steel H-piles, which penetrate conglomerate and dense sands and gravel quite well, do not behave well when they encounter cobbles and boulders or hard irregular rock. The flanges tend to tear and may curl up like spaghetti. There is often no easy way to verify their continuity.

The Norwegians have apparently been successful when founding piles through alluvium onto glacially smoothed rock, by incorporating a short (300 to 500 mm) dowel of high yield steel in the concrete pile tip to get a toehold in the rock.

For the Trident submarine wharf in Puget Sound, Washington, steel H-piles were embedded in the tips of the prestressed concrete piles in order to penetrate dense layers of overconsolidated silt. Both closed-end steel pipe piles and concrete piles penetrate rock fill embankment quite well, pushing the stones aside as they penetrate. With graded rock fills having maximum size stones up to 500 mm, little difficult is encountered in drilling and the pile location can usually be kept within 150 mm. Even when larger rocks must be penetrated, this can usually be accomplished, but tolerance in location will be greater, *e.g.* 300 to 400 mm for 1 m maximum size. The pile tip should be square, with chamfered edges. The pile must not be restrained once it has penetrated a meter or two. The old-fashioned practice of driving a spud and pulling it before the driving the permanent pile is counterproductive; when the spud is pulled, large stones are "sucked" into the hole.

Since many existing harbors have had several generations of wharf and pier construction, usually founded on closely spaced timber piles, it may be necessary to drive the new piles through the remnants. While, again, prior practice was to remove the existing piles, this is often not advisable. The old piles reinforce the slope; removal may lead to instability. In the case where a compact steel pile or prestressed concrete pile encounters an existing timber pile, it will usually deflect the timber pile out of its path.

When driving piles, they must not be unduly restrained, as this will induce unacceptably high bending stresses. During driving the pile is supported in the driving head and at the deck level. Once it is embedded in the soil, the gate at the deck level must be opened. Otherwise, if the tip of the pile drifts off because of encountering obstructions, there will be high bending at the deck level which will damage either the gate or the pile. This problem becomes more severe as the driving head approaches the deck level.

These warnings are especially applicable to the better piles. Due to the limited thickness of the typical deck, it may not be possible to get full fixity, but a high degree of partial fixity may be obtained by either embedment of the pile or by the use of reinforcing steel. Embedded steel piles should have shear keys or plates. They should be heavily confined by either reinforcing steel or plates. For a steel pile, when reinforcing steel is used, it can be welded in shear to a steel pile or embedded in concrete in the pipe pile.

For concrete piles the best connection is to form 50-mm holes in the head of the pile into which bars may be inserted and grouted.

After driving, the pile is cut off to grade; for a prestressed concrete pile, it should first be encircled by a 50-mm-deep cut made with a diamond or caborundum saw then broken down from above by a small air hammer. Without the precut, the concrete, being under stress, will spall badly.

In typical deep harbor situations, the piles are subject to continuous movement due to current and waves. In some cases pile heads may move up to 300 mm with the change in tidal currents. Batter piles, already stressed close to their bending capacity, may lean and crack. Therefore, a system must be implemented to stay the piles in position so they will be stable and the heads can be later incorporated into the deck.

Typically, timbers are used, bolted well below the future deck forms so that they will not interfere. They are secured by the friction development by the bolts. When batter piles are involved, the timbers may be supplemented by wire rope tie. Batter piles should be stayed and tied before the pile-driving rig releases its temporary support.

During the installation of piles concrete decks take very important role. The typical concrete deck usually consists of deep transverse cap girders incorporating the pile heads, working monolithically with a one-way longitudinal slab. If cranes will be employed, there will also be a longitudinal girder at or near the face and rear wall. This cast-in-place system is made continuous over the transverse caps.

Support for this during construction is obtained by transverse timber girts, accurately set to grade, locked to the piles by bolts. To increase the friction, high-friction pads may be placed on the timber-pile contact interface. Even then, the friction factors must be selected conservatively because the surfaces are wet. Timbers are used because of their strength in binding and because they will float when later removed. Gluelam timbers may be employed for heavier loads and longer spans. The critical design condition for these girts is usually horizontal shear adjacent to the pile.

The problem that develops is that the form supports must be set to exact grade and must be heavy timbers to carry the loads, whereas the stay timbers can be lighter and be set rapidly. Steel channels and even wide flange beams have also been used for longer spans.

When the loads from the fresh concrete of the deck, multiplied by an appropriate safety factor (1.5 to 2.0), are not able to developed by friction alone, than hangars are used over the top of the piles. The consequences of form slippage during concreting are severe, not only in relative to the lost concrete and bent reinforcement, but also possible injury to workers and pollution of the water by the wasted concrete.

The typically cantilevered deck overhang on the face of the wharf is especially difficult to support without sag under the weight of the concrete. The contractor has to choose between an extra temporary support pile or heavy cantilevered girts. Hangars are almost always essential on the outboard pile. Forms for the pile caps and girders are usually prefabricated as boxes, and may be set with cages of reinforcing steel inside.

Many modern heavy-duty container wharves are designed with a two-way continuous flat slab, hence no deep cap girder. Special shear-head reinforcement is required over the piles. The deck forms in this case are very simple, consisting of joists and plywood sheets. Nailing is only necessary to prevent dislocation by wind. This system greatly facilitates the future form removal from under the deck.

All deck forming should be designed with regards to its removal, since the removal is usually more costly and is definitely more dangerous to the workers than the initial placement. Form removal is usually performed by workers on a float using scaffolding. Life jackets are mandatory, and there should always be more than one worker present. To minimize the exposure of workers, temporary small holes can be formed in the deck so that the forms and girts may be lowered down into the water, then floated clear for retrieval and reuse.

Innovative alternative systems for forming have been developed, aimed at reducing the labor of forming and stripping. One of these utilizes precast concrete segments. First, lower half of the cap girder is constructed using light timber girts for temporary support. Then precast full-depth or half-depth precast slabs are set on the half-depth caps, and the whole made monolithic by cast-in-place concrete. This system requires that the cast-in-place half-depth cap be widened as necessary to correct for any mislocation of the piles.

For modern container terminals, designed for the berthing, unloading, and reloading of containers only, the fender systems are usually heavy rubber fender units, mounted on the face beam of the wharf, spanned by a steel beam and timber rubbing strips. The fender is supported by chains from the deck, both for gravity load and for the longitudinal displacement by friction from the ship. These latter chains are placed at a 45° to 60° horizontal angle.

More complex fender systems are required when the quay or wharf must handle barges and smaller ships. They must not only be fendered but prevented from getting underneath the dock. The outboard bearing pile must also be shielded from impact. Creosoted-timber fender piles were much used in the past, but their use is prohibited in many countries including the United States for environmental reasons. The U.S. Navy has therefore sponsored the development of prestressed concrete fender piles and also fiberglass fender piles. While more costly than the timber which these replace, then are able to accept more energy of impact and hence can be spaced more widely.

Hardwood timber rubbing strips are usually bolted to the face of the prestressed concrete fender piles. Such piles have given excellent service for over 30 years in the Arabian Gulf (Kuwait) and in the Singapore (Jurong). Steel concrete and timber panels, bolted to the face edge beam of the wharf, have also been used, but are very difficult to repair or replace after damage.

Single-wall steel sheet piles are the principle type, anchored back to deadmen at the waterline or slightly above. They develop the majority of lateral shear resistance by passive pressure on the below-ground length. Thus adequate penetration is essential, and design tip elevation must be achieved.

The Z-configuration pile is the most common, giving maximum bending strength while maintaining economy. For greater depths of water and for higher surcharge loads, H-sections or pipe-pile sections with interlocks are employed, often in combination with Z-piles.

Driving is resisted by tip bearing, by friction with the soil on both sides, and by friction in the interlocks. This latter item can be a major problem increasing with depth and resisting or even preventing attainment of the design penetration. Accuracy of the interlocks is controlled by the manufacturer. Binding of the interlocks is determined by the care with which the sheets are set. Hence, they should be set against a supporting wale with care to ensure verticality. The sheet piles have a designed spacing, usually giving 6 to 10 mm play in the interlock. Therefore, the location of each pile should be marked on the wale to prevent crowding or stretching. Z-piles in particular are easily rotated which not only increases the friction slightly but decreases the effective wall length. Lubricants such as greases are often employed but may not be fully effective due to contamination by sand grains. Hydrophobic sealants can be applied to the interlocks to make the sheet piles essentially watertight.

Since it is possible to set many sheet piles in a workday, there is a tendency to set them rapidly, without the required care. Then when it comes time to drive them, the friction increases and often gives the indication of having struck a buried log or other obstruction. On occasion it has proved necessary to extract the length of wall and reset the piles, a very costly and time-consuming operation.

The steel sheet pile wall, once set, is very exposed to the wind. Entire walls have been blown over, bent across the top wale. Since it is usually practicable to support from one side only, one out of every three or four sheets should be both tied and strutted. Driving of the sheet piles should proceed in echelon, so that no tip of an individual sheet is more than 1 m or at the most 1.5 m ahead of its neighbor. In the early phase of driving, where there is usually little resistance, there is a strong temptation to drive pilling down several meters beyond the adjacent piles to avoid changing the hammer. The problem arises due to the fact that sheet piles have low stiffness transversely and hence can be easily deflected by a relatively minor obstruction. Then when the adjoining pile is driven down, it is forced to deflect also, since the interlocking pile is firmly embedded in soil. The result is excessive driving friction and a high probability that the tips will not reach design tip elevation.

Vibratory hammers are especially effective for driving steel sheet piles. They temporarily liquefy the sands and silts along the faces of pile, which accounts for their ability to achieve rapid penetration. These hammers work best in cohesionless soils but are not effective in hardpan, or in layers of dense ash or peat. There, impact hammers must be used.

Driving sheet piles in the dense Arctic silts has proved impracticable unless the soil has been loosened by jets. On the West Dock at the ARCO facility on Prudhoe Bay, on the North Slope of Alaska, sheet piles met refusal at 10 m for both vibratory and impact hammers. Running a powerful jet down along the line of each sheet sufficiently loosed the slit that the piles were readily driven by the vibratory hammer. Interestingly, after a few days the disturbed slit regained much of its density.

A sheet pile bulkhead, being essentially a vertical wall, reflects incident waves, forming a clapotis, or standing waves, one-half wavelength in front of the wall. There is a strong upward flow of water immediately in front of the wall, which in the presence of even a moderate current, can lead to severe erosion. Use of a filter plus stone and riprap may be required. A filter fabric, properly overlapped and covered by rock, is preferable, although a properly graded stone filter may be used.

Prestressed concrete sheet piles are also used for bulkheads, especially those in shallow water, because of their superior durability. However, to achieve comparable bending moment to steel sheet piles, they must be relatively thick. They must be heavily prestressed. Interlocks are usually of the tongue-and-groove type, often with provision for grouting of the interlock after installation.

Typical dimensions of prestressed concrete sheet piles are 3000 × 1000 m. To overcome this large end-bearing area and the skin friction, jetting is usually necessary. The jet may be embedded in the pile, using thin steel pipe. Plastic pipe is not suitable, since driving requires use of an impact hammer. There is always the possibility of tensile rebound stresses producing a crack in the concrete pile which will also be reflected through the plastic, with the result that the jet water pressure now acts on the part or all of the cracked area, leading to hydraulic ram effects on the crack under every subsequent low, with washout of cement. Low-cycle fatigue will occur under as few as 20 to 50 blows.

Unlike steel sheet piles, prestressed concrete sheet piles are each driven to full penetration, one by one. To ensure that the subsequent pile will follow down close to the previous pile, the tip should be tapered an about a 45° angle, wedging the tip of the pile back against the first. The top can be held in alignment by a line from a hydraulic winch, with a roller to reduce friction.

Plastic interlocks have been used to prevent sand leaching through the joints as the tide alternately runs in and out. These have limited strength and cannot be counted on for structural purposes.

To provide a structural interlock, a half section of a flat steel sheet pile is embedded in each side of the prestressed concrete pile, and anchored by reinforcing bars welded to the sheet. The interface between the steel and the concrete will develop a high oxygen gradient, and hence be subject to corrosion at the line of embedment. Special attention should be paid to this corrosion susceptibility, *e.g.* by coating the line joint with epoxy or installing zinc anodes in the joint between the concrete sheet piles.

When driving sheet piles through rubble or debris, the integrity of the wall should be periodically verified by video or a diver.

The Study of River Structures

The River Structures include locks, low level dams, overflow structures and flood walls. Now let us study about the cellular cofferdams and Float-In Concrete Structures. By sheet pile cellular cofferdams a portion of the river, usually one third or the one fourth, is closed off and surrounded by multiple sheet pile cells. The site may have been dredged beforehand to remove unsuitable material. Each cell will be cellular, approximately the same diameter as the depth of water for which it is designed. Two or three ring wales are placed as a space frame, then held in place by temporary spud piles. Flat steel sheet piles are then set around the ring wale, taking care that they are truly vertical in both directions. To ensure proper spacing, the location of each sheet should be premarked on the top wale. In deep water and swift water, the lower wale may need to be located underwater. The sheets are then driven down with a vibratory hammer, working progressively around the ring, driving each in turn up to 1 m (maximum 1½ m) below the previous sheets.

Adjacent cells are connected by arcs, which engage "Y" sheets in the circular cells. Because the Y sheets are very stiff, they should always be set first in exact position and should always have their tip below those of the adjacent flat sheets. The sheet piles cells are then filled with coarse sand which is compacted by vibratory probes to a suitable density. This vibration builds up pore pressures which can lead to local liquefaction, especially if the cell fill consists of too fine sand. Vertical drains of gravel can be used to allow ready escape of the water. This also aids and speeds the process of desification.

Once all cells are filled with sand, a berm is constructed on the inner side to give added lateral support to the wall. This may be of rock which is free-draining. Most berms, however, are constructed of sand, with wells or well points to ensure against saturation by seepage. After dewatering, the slopes may be protected by a stone blanket over a filter fabric.

The cofferdam is now dewatered, and the structures built by conventional land methods. During this period, it is essential that a monitoring program be followed to reveal any scour along the cofferdam outer walls, so that if necessary, riprap can be placed. Water levels and pressures within the cells are also measured so that dewatering measures can be installed on a timely basis. Inclinometers check the cells for any tendency to lean.

If the river floods so that overtopping of the cofferdam is imminent, the cofferdam is intentionally flooded. Arrangements are incorporated at one or both ends of the new structure so that the end cells of the next stage cofferdam can be constructed. After all construction is completed, the cofferdam is flooded and the sheet piles are extracted, for reuse on the next stage cofferdam. There is always a certain amount of damage to sheet piles, so that only about 75 per cent or 80 per cent are usable in the next stage.

The setting of the flat section sheet piles poses difficulties when the water is deep, the current strong and the sheet piles are long. For initial picking, the piles should first be rotated onto their strong axis before being lifted. While lifting, the pile must be controlled by tag lines since otherwise wind will make them uncontrollable. Threading used to be done by hand. A person straddling the previously set sheets, using a portable chair, entered the interlocks. Today two alternatives are practicable and safe. One is to have an upper ring wale on which a person may stand and from which the sheet pile can be threaded. The other is by use of an automatic threader which rides up the interlock of the previously set sheet.

The current affects both the individual sheet as it is being slid down, and the partially set wall. That is why two or more levels of ring wales are necessary. The sheets should b secured against wind and wave action for both sides. In strong currents the use of a deflector shield may be necessary. This can be a fabricated shield which is secured temporarily by spud piles. It is located 5 m or so upstream. In lower currents, a deep draft barge, moored upstream, may be adequate. The upstream are of the sheet piles should be set first. Sheet piles in the downstream are and those parallel to the current

Figure 11.1: Temporary Structural Frame Mounted on Precast Shell, Olmsted Dam, Ohio River

should be temporarily tied back to the ring wales. In some extreme cases, a strong back may be needed to support the sheet pile while it is being lifted, up-ended, and entered and run down. The strong back is a wide-flange beam, with clamps that lock to the sheet pile until they are released mechanically or hydraulically.

Prior to arrival of segment, the site will have been dredged, rock sour protection placed and screeded to grade, and an anti-scour mattress placed. Piles are driven through hole in the mattress, guided by a light steel temporary template attached to the mattress. Leveling pads will be constructed to later support the prefabricated shell. These pads will consist of an underwater concrete cap, set accurately to grade on a cluster of piles. When the catamaran and its precast concrete shell arrive, the segment will be lowered onto the leveling pads, then pulled laterally into contact with the preceding shell. The shell is then filled with tremie concrete, placed through tremie pipe by gravity feed. These tremie pipes are guided and supported by the temporary frame attached to the shell. The connections between shell and frame are then disconnected, enabling it to be removed.

Complete sections of navigation structures such as locks and navigable pass dams are constructed in a construction basin or fabrication yard, then launched, towed to the site as self-floating structures, and installed by ballasting down. This process closely resembles that of prefabricated immersed tunnels (tubes). However, the differences are important.

First, the cross section of these structures may be irregular and require special ballasting in order to eliminate heel during the floating phase. The section is typically difficult to design and detail for resistance to hydrostatic pressures during installation. Second, unlike the tube which is set in a trench, the navigation structure extends above the bottom, up through the water. Positive cutoff has to be provided in order to prevent underflow. Third, because of the unequal water pressures and, in many locations, seismic considerations, lateral support has to be provided in the form of piling, or other means. For example, locks see many cycles of high global shear as they are filled and emptied. River structures see unbalanced water levels. Fourth, navigation structures often have exposed appendages such as movable gates attached which require special protection during tow and installation. Finally, navigation structures must usually be installed in close proximity to existing structures which have to be kept in service.

Prefabrication

Phase one of the construction is the prefabrication. Typically, this takes place in a construction basin. A thin layer of crushed rock is laid on the floor of the basin and covered with corrugated steel sheets or plastic. Plywood sheets covered with polyethylene have also been used. Note that whatever is used will stick to the bottom, held by adhesion and buoyancy. Therefore, if steel corrugated sheets or plywood are used without polyethylene, they should have ties to hold them to the bottom to prevent any possibility of their coming loose during final installation, where they might interfere with the set-down operations. If the base slab is to be later prestressed, the sheathing must slide to accommodate the shortening. More recently, the hard-finished concrete floor slab in the basin is coated with bond breaker and the new slab cast directly on it. Through plastic pipes spaced about 5 × 5 m, water is injected at low pressures for sustained periods of several hours; then the basin is flooded and the barge or segment rises clear of the bottom.

Construction then proceeds, often using precast concrete slabs, joined by cast-in-place joints. Construction joints must be properly prepared by water jet blasting to the depth of about 6 mm. Epoxy bonding compound may be used, applied to the joint face just before the concrete is placed. Forms for the joint must be grout-tight. Waterstops are not used; they impede proper concrete placement and just create a potential water path.

Reinforcing steel must go across the joint. While this is a matter for design, the constructor should be aware of the labor required in interweaving the steel. Accurate placement with staggered bars reduces the problem. Consolidation of concrete in deep, heavily reinforced joints may be added by preplacement of a tube. The vibrator is lowered down the tube, extending just below its tip, then both are gradually withdrawn as the concrete rises.

Roof slabs and other near-horizontal members are usually precast or else formed with stay-in-place forms. Corrugated steel flooring, plywood, and thin precast concrete slabs have been used. Precast concrete half-depth slabs, with protruding reinforcing ties, could also be used, designed to act compositely with the cast-in-place topping.

The final stage is bulkheading of the open ends, if any, in a manner similar to that for immersed tubes. Upon completion, flooding of the basin should be carried out, with an allowance of sufficient time for the water to fully penetrate the crushed rock base or under the slab before floatation occurs. The structure can then be winched out of the basin and towed to the site.

An alternative method is to cast the structure on a large barge or barges. The barges selected must be able to withstand the external hydrostatic pressure for their maximum submergence. Standard cargo barges may not be strong enough. If two barges are joined together, the connections must be designed to transfer the moment and shear. Tying across by beams on top and cables underneath, while suitable for moment and a·ial force, must be supplemented by shear plates. The connections must extend back far enough into the structure of each barge to prevent development of a week plane. Local loads may require internal posting of the decks.

The barge hulls, if floating during construction, will deflect as loads from concreting occur. Therefore, precast segments should be used, with essential all of them in place, before jointing. Alternatively, the barges may have been seated on a screeded sand base prior to concrete construction. This will allow concreting to proceed without special provisions for global deflection. However, even though the completed structure may provide shear strength to a barge combination during tow, it will generally not be adequate during launching, hence the previous provisions regarding joining of two barges still apply.

Launching from the barge can take place in two ways. The safest and best is by ballasting down while maintaining a level attitude. Stability during the several stages of ballasting must be verified, since the center of gravity is well above the center of buoyancy. The water plane of the barge provides stability during the initial ballasting; then there is usually a sudden reduction as the barge deck is impressed. Then the water plane of the concrete structure can provide the needed stability. However, if the structure has a profiled upper surface of variable depth, there may be some depth before float off which is unstable. For this stage, a temporary bulkhead may be required to maintain an adequate water plane.

When the concrete structure has sufficient water depth to float off, the barge or barges underneath must be separated from the structure by additional ballast. After separation, the barge below has little or no stability and is uncontrollable by itself. One way to solve this problem is to launch at a location where the depth is only a meter or so greater than that required, and the bottom is level. In this case the barge is over ballasted to sit on the bottom until the concrete structure has been towed clear. Then the barge can be deballasted by pumping from selected compartments and tipped up, using the end remaining in contact with the bottom to provide transverse stability.

A much better system is to provide steel hollow columns at the corners of the barge which provides sufficient moment of inertia through the water plane for stability at all stages. This tipping down process can also be used for the launching, if the water depth is less than one third of the barge length. The critical situation here is transverse stability as the deck of the barge goes underwater and the combined barge-structure unit tends to roll sidewise. However, this inherently operation has been successfully performed if the concrete structure has sufficient size in plan to provide water plane stability during the immersion stages.

Installation

Installation at site requires both horizontal and vertical control. Horizontal control is usually provided by moorings, running to preplaced anchors. Winches are mounted on the unit being placed. Typically, a temporary steel frame is left on the top of the preceding unit. It extends up above water and thus furnishes a convenient reference for locating the new segment relative to the preceding one. Except in unusually deep water, this is more practicable than locating the new segment by sophisticated electronic, differential GPS, and underwater sonics.

The new unit can be guided to exact position by tapered wedges or horn guides. Alternatively, it can be lowered to final elevation, but offset 100 to 150 mm, then, while it is still buoyant, drifted laterally to contact.

As with precast shells leveling pads will be pre-installed. Practice differs as to wheather two of these seats should be extensions from the previous segment, with one or two independent pads located at the leading end, or three or four independent pads. The first method ensures correct relative elevation, the second eliminates overload on the previous segment. The determining factors are the foundation capacity and the allowable underwater span length of the new segment. The method for construction of leveling pads has evolved through a long series of projects.

In Norway, with rock outcrops on which the segment must be seated, precast concrete slabs are lowered in exact lateral position, to rest on the seafloor at a slightly lower elevation (-150mm) than final requirements. The slab will have flexible skirts or, in shallow water, the diver will place sandbags to seal underneath. Grout will be pumped under the slab. Once it has set, an accurate measure will be made of three points on the slab. Precast concrete "shim" slabs will be then placed to raise the top to the exact grade.

In The Netherlands, similar pre-cast slabs and shims are used, but here the slab is initially seated on a crushed rock base. For the Prince Edward Island Bridge, a three-legged jacket was set on the dredged rock seafloor. A central tube was accurately positioned and plumbed to vertical. Tremie concert was then placed to a slightly low position. The surface was cleaned by jet and airlift. After measurement, a precast shim slab was placed. For the Oresund

Bridge, the support pads were varied from 1.5 m square to 3.0 m square. They were 150 mm thick. The pads were attached to a three-legged tower (jacket). The jacket was precisely located by DGPS. Jacks in the leg of the tower adjusted the height of the pads. Sand-filled sacks were used to seal underneath and grout was injected. For the Olmstead Dam, leveling pads will be supported on four pre-driven steel piles whereas for the Monongahel Dam #2 they will be supported by drilled-in-shafts.

Because the leveling pads are "hard points" and furnish support during the subsequent placement of tremie concrete, they may experience greater loads than the remaining piles or shafts. Similarly, the segment will undergo higher bending during the subsequent infill ballasting and concreting. Both temporary and residual stresses and loads need to be calculated and considered. To overcome this matter of hard points, when the structure is to finally rest on a sand or crushed rock base which is slightly compressible, the pads may include a layer of wood or polyurethane which will yield during infill and transfer the load to obtain a more uniform distribution. Flat jacks may be fitted, either on the top of the leveling pad or on the underside of the segment, permitting adjustments of load. When final seating has been obtained, they are pumped full of grout.

Now let us study about underfill. Underfill may be sand, concrete, or grout. Sand underflow systems have been used in Denmark and The Netherlands, for which liquefied sand-water slurry flows down and automatically fills underneath by channeling. Underbase grouts have been used, varying from a cement-sand grout to foamed cement grout, depending on the requirements and the depth of fill required. Heat of hydration must be considered; pure cement grouts or cement-bentonite grouts can get excessively hot, even over the boiling point. They may damage the segment's base slab.

Concrete underbase fill, with aggregates, gives higher strength, high modulus, and reduced heat of hydration. Usually small size coarse aggregate, 8 to 10 mm, is used. Limestone powder and fly ash are means for reducing heat and modulus, minimizing the potential for hard points under the base slab. Underbase fill must be placed under low pressure to avoid excessive pressure, both locally and in global uplift. Gravity feed should be used in preference to pumping, since friction head loss in uncertain and the tendency is to develop too high pressures in order to ensure flow. The edge and the joints must be sealed to prevent flow out to the water. The current tends to suck the grout or concrete out. Sandbags and grout-filled bags have been used to seal, also expand metal screens with filter fabric attached.

Filling completely underneath a horizontal slab requires use of a self-leveling "flowing" concrete or grout. It also requires a positive head to overcome friction and to displace water and laitance. An antiwashout admixture based on welan gum is a valuable aid in obtaining the required properties and in minimizing or eliminating bleed. To ensure shear transfer, corrugation or horizontal shear keys can be provided, but the location of discharge and venting holes needs to be properly planned. Vents are required to expel the water and any laitance or foam.

To transfer tension requires that steel bars, strands, or shapes extend down from the base slab of the segment. This, of course, presents difficulties in the prefabrication stage, requiring that the base slab be supported on a raised soffit. It also increases the draft during tow. Care must be taken at all stages to avoid crushing these or, worse, pushing them through the base slab, due to accidental grounding on an obstruction. For these reasons the use of semiflexible strands where structurally adequate is preferable. The planned construction for the Inner Harbor Ship Canal Lock in New Orleans will require the transfer of tension and shear through the tremie concrete underbase fill.

The Box Caissons

Box caissons are prefabricated of steel or reinforced and prestressed concrete. Large caissons, for which the shells exceed available lifting capacity, *e.g.* more than 3000 to 8000 tons, may be constructed in a basin or graving dock. When complete, the dock is flooded and the caisson floated. The dike or gate is opened and the structure floated to its location where it is positioned, seated down by ballasting onto its prepared base, and permanently weighted down by underwater concrete and stone or sand fill.

Stability is maintained hydrodynamically by always keeping a positive metacentric height, $\overline{GM}$, supplemented if necessary by attached buoyancy tanks or support from external vessels. Particular care has to be taken at abrupt changes in cross section, *e.g.* at the transition from a large footing block to shaft. The foundation for the caisson must be level, to a close tolerance, and sufficiently dense so as not to settle differentially, not to liquefy due to rocking of the caisson under wave action, and not to liquefy under earthquake.

Initially the unsuitable soils are removed by dredging. Usually a clamshell, hydraulic backhoe, or ladder dredge is used. Soft sediments remaining must be removed by suction, *e.g.* by use of an airlift. The surface must be sufficiently clean that a layer of weak clay does not exist across the foundation, but limited amounts of soft silts may be left in isolated pockets. If the soil is sand, densification may be undertaken using heavy vibratory probes. Sand or small stones may be fed down during the operation. Air pressure and vibration causes the sand and stone to flow out to create a dense foundation which is free draining.

A stone bed is then placed. This bed is constructed of strata, about 1.0 m deep, which can then be compacted by a large and powerful vibratory scared. Finally, the surface is screeded to exact elevation, usually ±20 mm, using a 300 mm layer of 12 to 20 mm stone. The screed is usually supported on preplaced rail beams. A traveling hopper is fed by a tremie pipe or pipes from the surface. The stone is then distributed laterally and struck off to grade by a blade on the trailing edge of the hopper. In some recent cases, the distribution and striking off is accomplished by use of a continuous flight auger, riding horizontally on the rail beams.

For the Honshu-Shikoku Bridges, the foundation was excavated to bedrock or hardpans, then leveled carefully to ensure proper seating of the peripheral edges of the caisson. Compressible seals were used under the outside bearing surfaces.

If there has been any appreciable delay after final screeding of gravel, then prior to positioning the caisson, a bottom survey should be made, either by acoustic depth finder from the surface or by side scan sonar to verify that there are no obstructions.

The caisson is now floated into position. Although GPS is used for general control, a laser range is very valuable for initial positioning due to the fact that the amount of correction needed and the rate is readily apparent to those controlling the position. Theodolites are used for initial precise control.

Moorings are used to effect final positioning to the required tolerances. In order to facilitate corrections, lines should be led out orthogonally, each subject to separate control by a winch. This typically requires six to eight lines. Attempts to reduce the number by running lines at 45° are usually counterproductive since taking in on one line moves the structure on two axes and rotates it. Preset mooring buoys are installed, each set with three anchors. Once the caisson has entered the area, lines are run from winches mounted on the caisson, initially leading down to fairleads attached to the sides at the approximate center of rotation, then led out to the buoys. The caisson is then ballasted down onto its prepared base. The rate of ballasting down over the last 2 m should be slowed, in order to let the trapped water escape. During this final set down the current will accelerate underneath the base, tending to pull the caisson down while at the same time scouring underneath. In fact, in very loose sands, the bottom may drop progressively as the caisson is lowered. This may require prestabilization of the foundation by willow mats with stone or with their modern replacement, filter fabric plus stone.

Immediately after seating, position, orientation, elevation, and level are checked to ensure they are within tolerance. If one parameter slightly exceed tolerance, an inquiry should be made to the design engineer, to see if the caisson can be accepted as is with perhaps minor changes to the superstructure, since the act of de-ballasting and re-positioning often causes damage to the prepared base.

Large box caissons for bridge piers which are being floated into position and then submerged are usually of prismatic shapes and hence have a substantial waterline plane through the whole seating process, which ensures stability at all stages. Conversely, those which have a radical change in waterline plane, *e.g.* where the top of the footing block is immersed and only a smaller shaft extends through the water plane, usually require some external means of providing stability. One reliable means is the attachment of vertical buoyancy columns or tanks, rigidly connected to the caissons. Post-tensioning the column from the top, through a steel pipe sleeve, and anchoring in the caisson is a useful solution because it permits subsequent release from above water. Another often used means is to supply stability from a large crane barge, since a lift on long slings can create a righting moment $Pl \sin \theta$, where P = the lift force, l = the length of the slings, and θ = the angle of heel; such a lift effectively raises the $\overline{GM}$ by $P \times l/\theta$ delta where "delta" is the displacement.

Once the caisson is in position and fully ballasted, grouting underneath should be performed promptly. It is important to prevent runout of the grout under the edge of the caisson, so the caisson base is usually provided with short skirts to retain the grout. A grout with low heat, antibleed, and antiwashout characteristics is desirable. The grout should have low modulus, hence low strength, so that local deformation can be accommodated. Vents must be provided in the base of the caisson to relieve pressure, and the grout injection pressure carefully controlled to avoid raising the caisson. During this operation, the elevations of the corners should be continuously monitored.

The two foundation piles for the Ma Wan Tower of the Tsing Ma Bridge in Hong Kong were prefabricated box caissons, cast on a large submersible barge, of the type used for ocean transport of dredges and jack-up drilling rigs. The barge, with the two prefabricated box caissons abroad, was towed to a site where the water depth was just adequate for the launching operation. The bottom was sand and approximately level. The site was carefully surveyed, both acoustically and by underwater video, to ensure that there were no boulders or obstructions. Both caissons were securely chained to the barge in both transverse and longitudinal directions to prevent a premature launching. The barge was then ballasted down to the point that the caissons were close to being buoyant. The barge was then given a 3° trim down-by-the-stern, the rear caisson cut loose, so that it slid off the stern. The barge was now freed of half its load and began to float up. This was the most critical point for stability. In this case, the submersible barge performed according to calculations and remained stable.

The process was then repeated and the second caisson was launched. The two were now floated to the site. It was necessary to move into position at high tide to have adequate under clearance. Each caisson was moored and ballasted down onto seating pads which had been built to exact grade. In this case, the "prepared" foundations consisted of bedrock which had been drilled and balasted, then excavated. Although grate care had been taken in the balasting operation, *e.g.* close spacing, minimum charges, and minimum overdepth blasting, the rock had many natural and man-made fractures. Plastic tubes were inserted in the prefabricated footing blocks so that when they were later concreted, drilling and grouting of the fractures could more easily be carried out. The bedrock was thoroughly cleaned by airlift. Sealing of the edges was difficult, requiring diver packing by grout-filled bags. Then tremie concrete was placed to fill the box. To prevent runout underneath, a stiff mix was used. Sandbags were placed to fill the gaps at the edges. The box had a slopping half-roof. Due to the sniff concrete, as well as bleed and laitance, a void arose under the roof. This was filled by supplemental grouting. Use of a concrete mix containing an antiwashout admixture and higher slump would probably have eliminated the void. Flow out under the edges might have been more easily prevented by crushable or flexible skirts, *e.g.* canvas with heavy chain sewn into the bottom.

In the case of the earlier Honshu-Shikoku bridges across the Inland Sea, the steel box caissons had preplaced grout pipe installed. Once the box caisson had been founded, it was filled with coarse aggregate, from which all fines had been screened out. An algae inhibitor was added to the water to prevent the growth of slime on the aggregate. Then grout was pumped through pipes, starting with those discharging at the lower levels and progressing upward. Indicater tubes were used to monitor the rise of the grout. However, subsequent corings indicated significant zones which remained ungrouted. It is believed that these were due to trapping of bleed water. On the most recent and largest of the Honshu-Shikoku bridges, that across the Akashi Strait, tremie concrete was used in lieu of grout-intruded aggregate. A special low heat of hydration mix was employed, incorporating limestone powder, fly ash, and blast furnace slag.

Scour protecting must be placed as soon as possible to prevent erosive scour of the prepared bed. Shortly after installation, one bridge pier box caisson on the Øresund Bridge, between Denmark and Sweden, was seriously undermined by scour from the joint action of waves and current. Extensive grouting was required to restore the foundation.

The process described earlier for grouting under the base to fill completely any gap between it and the stone bed was used on the caisson piers for the Eastern Bridge across the Great Belt of Denmark. The caisson bases were configured with three or four integral bearing pads at the corners, typically 300 to 500 mm thick, to ensure that a proper gap was maintained for grouting. Alternatively, bearing pads may be preset. For the 44 caisson piers of the Confederation Bridge in eastern Canada, three bearing pads were preconstructed by placing tremie concrete to the exact grade, using a small bottom-founded jacket. Then after the caisson was set on them, the bearing space was filled with a flowable tremie concrete containing silica fume to ensure cohesiveness. However, on the Western Bridge across the Great Belt, the 66 box caissons were set directly on the stone bed. This concept requires that the base be adequately reinforced to span any gap between high points (hard points) and that the screeding be carried out to grate accuracy.

On the Second Severn Bridge, prefabricated box caissons with open bottoms were loaded by transporters onto barges, then towed to the site at high slack water and picked by a pair of heavy-duty, high-capacity cranes supported on a large jack-up barge. After setting, they were filled with tremie concrete. Initially; difficulty was experienced in sealing the caisson edges against the effect of the very strong tidal currents, which led to extensive washout. Eventual solution was found by placing closely meshed screens around the perimeter and using a thixotropic antiwashout admixture in the tremie concrete.

A similar problem had occurred many years earlier on the Columbia River Bridge at Astoria. Thereafter the box caisson shells had been placed, the river currents augmented by the outgoing tide scoured the sand from under the base. When the tremie concrete was placed, it flowed out and was washed away by the current. This problem was only remedied by placing a filter course and scour protection around the pier shell, filling above the bottom of the shell and followed by extensive under base grouting to fill the voids which had been formed during the initial concrete placement.

The foundation piers for the Rion-Antirion Bridge in Greece must cope with deep water, soft soils, and a major active earthquake fault in mid-channel. The piers are constructed as gravity-base caissons, founded in the soft clays which are improved by deep-cement mixing to provide adequate bearing and shear resistance.

Where heavy-lifting equipment can be made available, the complete pier is prefabricated, either in a special job-site plant or on a quay (bulkhead wharf) of adequate capacity. Then it is lifted by crane barge or catamaran and carried to the site, to be set in place on the prepared stone bed. As noted earlier, many box caissons are configured as a submerged footing block, supporting the lower portion of a shaft which then extends through the water plane. If the shaft is not mounted until after the footing block has been fixed in the base, then there are two solutions. In the first, a short box cofferdam of steel piles in temporarily affixed to the top of the footing block caisson, to permit dewatering and construction of the shaft in the dry. In the second solution, the shaft is precast as a hollow shell. It is then set over projecting reinforcing bars, sealed to the top of the footing block by gaskets, and filled with tremie concrete. The projecting bars may have mechanical heads in order to gain positive anchorage.

There are many variations for making a full-moment connection between a precast shaft and the footing block. For the Prince Edward Island piers, they used the mortise-and-tendon concept, with the hollow shaft fitting closely over a 1-m high tendon. The annular space between was grouted. On the Richmond-San Rafael Bridge, and again on the San Mateo-Hayward Bridge, a ring of heavy reinforcing bars extended downward 2 m into a recess on the footing block. The void and the lower part of the shaft were then filled with tremie concrete.

The early success with the underwater precast shaft to the footing connections had led to the current practice of making the above-water shafts by means of precast segments, each joined to the other by setting each upper hollow shafts segment over reinforcing bars projecting from the previous segment. The hollow in the shaft segments does not have to be of constant cross section for full height, but can be principally a pocket of sufficient depth to accommodate the projecting bars, with a smaller vertical hole for placing the tremie concrete. This concrete should use 10 mm maximum size coarse aggregate, a relatively high cement factor, and silica fume, the latter to eliminate bleed, gain strength rapidly, and to enhance bond.

Pile-Supported Box Caissons

In this concept, the prefabricated concrete shell is seated over pre-installed piling. The box is then filled with tremie concrete, locking the box to the piles. In many applications, the box has no bottom slab, so must be transported on a barge, to be set by a crane barge, or must be transported while suspended from a crane barge or catamaran, then lowered into position. The piles are best driven through a template, such as a grid of steel beams or a precast concrete flat slab with appropriate holes, in order to ensure that they are correctly positioned and particularly to ensure that the pile heads all lie within the periphery of the box.

On the other bridges, such as the Narragansett Bay Bridge in Rhode Island, the bottom "slab" was a heavy mat of timbers and steel beams, similarly serving both as a template and support.

For a current bridge across the Kennebec River in Marine, large-diameter drilled shafts are being first installed through a template. These vertical shafts have permanent steel shell casings. The drilled shaft-tubular piles are cut off about 1.5 m above the bottom of the footing block. Then precast boxes will be floated in. Each box has four externally fitted spuds (1 m diameter steel tubulars) which can be dropped to hold the box in position while it is being lowered over the piles. The base slab of the box has large circular sockets which will fit over the tubular piles. When the box is at the proper elevation, it is fixed to the spuds. The annuli around the piles are sealed and grouted with containing silica fume. Then after the grout has gained strength, a tremie concrete seal coarse is placed. The box, while floating as a prefabricated box, has a temporary cofferdam attached on top, which will not only give stability during immersion but enable dewatering after the tremie concrete seal coarse has been placed, and completion of the connection of the caisson footing block to the drilled shafts.

On the first Benicia-Martinez Bridge, prefabricated box caissons were floated into place and moored. The box had sockets for the 2-m diameter driven and drilled piles. These sockets were covered during the flotation mode. One by one they were uncovered and the socket used as a template for setting and drilling of the piles to bedrock. After four piles had been driven, the box was ballasted to grade and then grouted to the piles to fix the elevation of the box. Then the remaining piles were driven. All piles were then socketed into the shale bedrock. A reinforcing steel cage was placed and whole pile and socket filled with tremie concrete. The annuli between the piles and the base slab of the box were then grouted and the pockets in the footing block were filled with concrete to transfer shear.

This project, completed in the 1960-70 decade, has been performed well. Recently, however, it was reanalyzed for a much large earthquake using up-to-date seismic input projection and analyses. As a result, additional driven-and-drilled tubular piles will be installed, the remaining voids in the box filled with concrete, and the entire box will be post-tensioned in the two orthogonal directions to lock the piles to the footing block.

A similar process will be used on the Second Benicia-Martinez Bridge, and is under consideration for the San Francisco-Oakland Bay Eastern Replacement Bridge. On these bridges, the contractor will be given an option to construct the box above water, on temporary falsework, then lower the caisson into place. This will be an alternative to off-site prefabrication and float in. As before, once the caisson is at the proper elevation, the annuli around the piles will be grouted and a tremie concrete seal course placed. A heavy cage of reinforcing steel will have been installed, either before lowering or after placement of the seal course and dewatering. Temporary small cofferdam boxes will be affixed to enable placement and stressing of horizontal post-tensioning tendons, as well as their subsequent sealing and grouting.

The Installation of L.D. Steel Tubular Piles

In installating offshore piles, their positioning is obtained by typical threading sleeves in the jacket, whereas, for bridge piers, jackets are not normally employed for a variety of reasons including the generally limited depth of water in which bridge piers are usually founded. Nevertheless, pile location and verticality or inclination are even more important for bridge piers than for offshore platforms. This is particularly true for large-diameter tubular piles.

Temporary templates are usually employed for the purpose of accurate initial setting where the bottom has a slope or where inclined better piles are to be installed and special procedures must be used, since once a pile has penetrated two or three diameters, it is impossible to correct it if it starts to drift off. Further, a pile which is drifting out of plumb or position can move for template laterally. Therefore, in such situations, it is best to partially drive one or more vertical piles in the cluster first, to secure the template location. Then the remaining piles may be driven. Finally, the four initial piles can be spliced and driven. The sequence should be selected so that after one battle pile is driven, the opposing batter pile is driven next. The template should be ideally provide support at two elevations.

Pile hammer selection and use follow closely the practice on offshore platforms. This matter is presented in detail in chapter 8. To obtain penetration in dense soils, the hammer should if practicable be sized to do this by driving alone. This may mean the provision of very large pile hammers along with the crane equipment necessary to handle them. However, it may not always be feasible to obtain the design penetration by driving alone. Jetting, clean-out, and drilling ahead are often required.

Unlike offshore platform installations, the need to develop tension and lateral capacity and to minimize settlements may lead to serve restrictions on installation methods. Thus external jetting may be prohibited in order to maintain the effective lateral support. Cleaning out helps in obtaining penetration by reducing the internal skin friction, but this in turn reduces the effective end plugging. This then allows more soil to move laterally until end plugging is regained, with a consequent reduction in the effective soil friction on the sides of the pile. Internal jetting to breakup the plug has a similar but leaser effect. In extreme cases, the clean-out and jetting processes may allow a temporary void to form beneath the pile, with a consequential permanent loosening of the lateral soil confinement.

Soils near and below the tip which have been loosened by the process described above allow greater settlement under loads than where the penetration has been obtained by driving alone. Thus the common bridge design practice of requiring clean out and installation of a concrete plug may be counterproductive. This practice is not used for offshore platform piling. The designs for several recent bridges provide for clean out and concreting of only upper 10 to 20 m.

Where piles are jetted to aid penetration in dense sand, much of the original capacity of the sands can be restored by continued driving after the jetting has stopped. From 50 to 200 blows of the hammer may fail to develop any significant additional penetration but will compact and densify the sands, internally and externally. Some archaic piling specifications in bridge practice attempted to attain this effect by requiring a penetration after the jetting ceases of 1 to 2 m, but with large-diameter piling, this is often impracticable to achieve.

One method that has successfully applied in order to limit settlements under high loads is to clean out the pile to within a few meters of the tip (typically one diameter) and then place a concrete plug of sufficient length to develop load transfer in bond. Beneath the plug, flat jacks or similar are installed. After the plug has gained strength, the jacks are pressure-grouted. A variant on the process is just to the grout the space without flat jacks, controlling the pressure so as not to cause run-out around the tip or fracture the formation. The body of the pile is then filled with sand or concrete.

When driven tubular piles require a socket beneath their tip in order to achieve the required capacities, then obviously the pile will have to be cleaned out to its tip. Cleaning out by mechanical means such as grab will cause less disturbance of the external soils than a jet and airlift, at least over the last several diameters. In all cases of clean-out the pile must be kept full of water to near the external water level. If the water inside is lowered, *e.g.* by airlift, a sudden inflow under the tip may ensure; if raised more than a few meters, *e.g.* by jet water, a sudden outrush may occur, causing piping along the side of the pile.

This regulation of pressures can be provided by pumping in continuously as an airlift is operated, or pumping out continuously as the jets are operated. Where design considerations permit, holes in the pile wall below water may provide equalization. Experience in bridge foundations has shown many cases where it proved difficult to maintain this head while operating an airlift. Consequently, the use of the grab bucket is now being increasingly employed or specified. It is as especially appropriate method where contaminated soils are involved or where there is restriction on discharge of sediments into the water. Even the grab bucket must be designed so that the water can equalize above and below while the bucket is being lowered and especially while being raised. For this purpose the garb bucket should have holes or provide spaces around its periphery. Alternatively, an augur or bucket drill is employed.

Due to limitations on crane boom reach and length as well as lifting capacity, the tubular piles may have to be spliced during driving. This operation is not only very critical to the integrity of the pile, but is also very costly in time and money. Further, the stopping of a pile section with its tip in dense material may lead to setup and difficulty in getting the pile penetration restarted. Hence, the location of pile splices needs to be carefully considered.

As noted, steel tubular piles for bridges may be made to act in composite action by filling all or a portion with concrete. Filling both increases the stiffness and prevents local buckling of the pile under seismic or impact. Although bond is often adequate to transfer the shear from the steel pile to the concrete core, near the head, shear connectors are required. Induction-welded studs are often specified but have the problem of interference with drilling and clean-out operations if these become necessary. Weld beads are sometimes employed but the most reliable is the use of spiral or circular bars such as 10 mm flat bars, welded around the inner circumference with fillet welds. The shear connectors should be continued over a sufficient length to transfer the shear due to high moment just below the pile head, usually one to one and one half diameters. Similar shear connectors are also often used to lock a concrete tip plug in place where such a plug has been specified.

An interesting concept was developed by Caltrans for two additional support piers for seismic retrofit of the San Francisco-Oakland Bay Bridge. The piles had to be installed through 20 m of water, then 35 m of very soft mud, and 40 m of clay, to found on bedrock. The rock was weathered and fractured to an additional depth of 10 to 15 m. Very high lateral seismic accelerations (in excess of 1.5 g) had to be resisted in essentially elastic action. The requisite combination of stiffness and strength was obtained by a double steel shell. The outer shell was 4 m diameter and the inner shell 3 m. The outer shell was to be driven first and seated in the rock. Then a socket, of slightly less diameter, would be drilled through the fractured rock and on into sound rock a sufficient distance to develop fixity. The inner steel tubular was then set and grouted to the rock. Tremie concrete was placed in the inner pile up above the rock line. The annulus between the two tubulars was now filled with tremie concrete, leaving the inner pile empty up to a top concrete plug.

For bridge pier piling, with long design life, corrosion protection is of increased importance. The use of an additional sacrificial wall thickness (typically 6 to 12 mm increase), or the provision of sacrificial anode cathodic protection below water, especially at the mudline, and epoxy or other protective coating in the zone near the waterline are the most common solutions.

Steel tubular piles, 2.5 m diameter, are planned for the new San Francisco-Oakland East Bay Replacement Bridge, as well as the

Woodrow Wilson Bridge across the Potomac River. Lengths will be up to 90 m, penetrating deep bay mud and plastic clays to dense sandy clay and hard overconsolidated clay. Where the water is deep, batters up to 1 to 6 will be employed, whereas at shallower sites or where upper sands are present, only vertical piles will be employed, the objective being to obtain maximum ductility in earthquake. In both cases, piles will be positioned by templates.

Steel tubular piles will also be employed to stiffen the response of the piers of the Richmond-San Rafael and San Mateo-Hayward Bridges under a projected strong earthquake.

The largest-diameter piles have generally been manufactured as vertically cast segments, then tipped to the horizontal and assembled by post-tensioning. Joints have been cast-in-place concrete. The large piles are transported either on barges or suspended between two pontoons. This latter method make it less demanding on the pile when picking, since the lower half can be partially supported by its buoyancy during up-ending. In sandy soils, these piles have been sunk in place by a combination of their own weight, plus pull-down forces from the crane barge, plus jetting and internal excavation by a dredge head.

In the case of King Fahd Causeway Bridges, holes were predrilled, using temporary casing, then the piles were seated, grouted in place, and plugged by tremie concrete. "Piles," 10 m in diameter, were employed for the Yokohama and Hiroshima Bridges in Japan. Those for the Yokohama Bridge were sunk by progressive drilling and under-reaming while those for Hiroshima were sunk by dredging and internal clean-out.

Circumferential confinement must be adequate to prevent vertical cracks due to thermal and structural action. Due to the prestress, inclined and radial stresses all combine to produce vertically oriented cracks between the longitudinal strands. Experience has shown that the circumferential steel must be adequate to match the tensile capacity of the concrete cross section without exceeding yield stress in the steel. This typically requires a steel area of 0.8 to 1 per cent.

One of the major advantages of these large-diameter prestressed concrete piles is their durability and corrosion resistance. This property enables them to be extended through the water column and splash zone. However, durability can only be assured by adopting methods appropriate to the environmental exposure. Below water level, where oxygen is limited, durability can be achieved by using a dense impermeable concrete, *e.g.* a concrete mix including silica fume and an adequate cover over the outer steel. In this below-water-zone, marine life will attach itself. Dense impermeable concrete is normally impervious to marine borers. However, in the Arabian Gulf, marine bores can eat through the cement coating into the limestone beneath and quickly bore to a depth of 50 mm or more. In locations such as these, only silica (igneous) aggregates should be employed. On the Ju'Aymah trestle, where high-grade limestone aggregates were used, the piles were quickly riddled with holes. Repair was made by removing the barnacles and other growth, filling the holes with epoxy paste, and coating the pile with hydrophobic epoxy, which could be placed in water. With the knowledge gained from that experience, on the King Fahd Causeway Bridges igneous rock was used for both coarse and fine aggregates.

As the seafloor, moving sand may abrade the pile surfaces. Generally this is not significant except in the surf zone. Again the use of dense, impermeable concrete, obtained by the use of silica fume is the best answer.

In the splash zone, there are many aggressive forces to be dealt with. The primary problem is that of corrosion of the reinforcing steel, since there is available oxygen from the air, chlorides from the sea, and water to provide the electrolyte. Although impermeable concrete and adequate cover are the principal protection, this zone is subject to cycle wetting and drying as well as cyclic thermal gradients, all of which combines to increase the permeability. Epoxy coating of reinforcing bars has been tried but is not generally satisfactory due to the loss of adhesive bond and delamination in seawater. The concrete cover tends to spall during driving and under bending forces. External epoxy coating of the concrete surface of the piles in the splash zone was employed on the King Fahd Causeway Bridge Piles. The coating extended up to the bridge superstructures. Three coats were applied. After 10 years, detailed inspection showed no chloride penetration whatsoever. The epoxy coating extended up to +4 m. Examination showed considerable chloride penetration into the concrete from +4 to +6 m. It appears that the splash zone needs to be extended well above normal high water and waves. The outer epoxy coating is, of course, subject to considerable UV deterioration and must therefore be re-applied every 10 to 20 years.

In locations subject to sustained freezing air during winter, yet here the tide periodically causes a rise and fall of the water, the concrete is exposed to severe freeze-thaw attack, along with thermal gradients and wetting and drying. Solid concrete piles with adequate spiral reinforcement have proved durable, but hollow core piles have suffered vertical cracks which open wider over time due to ice-jacking. Freeze- thaw attacks the cracks surfaces. The exposed circumferential steel then corrodes. The primary means of prevention is to incorporate air entrainment in the concrete and to use substantial circumferential steel, so that under the combination of thermal change and wetting and drying, the reinforcing steel is never stressed beyond yield. High bond is desirable to minimize crack width. Epoxy coating of the reinforcing bars is therefore not appropriate. In the Chesapeake Bay, application of an external epoxy containing glass spheres, coating the surface of the concrete piles through the splash zone, has proved to be an acceptable solution.

Now let us study about Large Diameter Drilled Shafts. This concept is a very economical and attractive method for constructing the pile foundations. It is efficient when the lateral loadings are low and where the soils are competent at a few meters under the mudline.

Casing installed by vibration or impact driving is normally required in order to exclude the water and prevent run-in of the loose surficial soils. Drilling is carried out with either full-diameter rotary drills, such as a pile-top drill, or with down-the-hole drills. The water level in the casing is maintained at the external water level or slightly (0 to 2 m) above. Rotary casing drills are also employed, for which the casing tip is fitted with teeth or bits, so that rotating the casing cuts into soft rock.

When the shaft is constructed through sands or silts or shales which degrade on contact with water, slurry is utilized to help hold the hole open. In the past, bentonite slurry was much employed when seawater alone was insufficient. Although relatively inexpensive in itself, many countries prohibit its discharge into water alongside, so that it has to be contained for reuse on adjacent piles. More serious, in the case of silts, mudstone, and shales which are impenetrable, the bentonite degrades the bond with concrete.

Attempts to wash it off by circulating water immediately prior to concreting sometimes leads to fall-in of the sides.

Recently, polymer slurries have been developed which biodegrade over a short period of time and hence are environmentally acceptable. These actually enhance the bond of concrete to the walls of the shaft. Although relatively expensive as a material, their use overwater reduces the operating costs and produces more reliable performance.

Since the drills operator beyond the tip of the casing, they must be held in axial alignment by centralizers on the drill stem. Special care has to be taken when the shaft is inclined (battered). A stiffer drill stem and more frequent centralizers are partial solutions to the matter of droop.

As with all drilling operations, care must be taken to prevent loss of circulation, in which the drilling fluid under its excess pressures escapes under the casing. This last is a common problem when constructing a drilled shaft through loose sands to an interface with rock and can usually be prevented by the following procedure.

Drive casing and seat it as practicable in the hard or tight formation, but without driving so hard as to damage the tip.

Clean out the pile with a balanced water head.

Use of churn drill or down-the-hole drill to drill ahead for 1 m or so. The churn drill is very good in rock because it tends to wander slightly and hence develops a hole slightly larger than the casing O.D.

Reseat the casing by driving.

Drill the socket.

Belled shafts have been extensively used on land to take advantage of the high end-bearing capacity of particular strata of limited thickness. However, it was only with the advent of driven and drilled tubulars in the North Sea that the concept was used for overwater structures. Because of the potentially increased problems of fall-in and the consequences of a failure, the "casing" was actually a driven tubular pile. The casing was cleaned out and the belling tool run on the drill string, using centralizers to ensure that the bell was axially aligned and held with sufficient stiffness to prevent wander.

The first task is to drill a hole of the I.D. of the tubular to about 1 m below the final grade of the bell, to act as a sump to collect cuttings. Then the arm of the belling tool is gradually expanded in stages until the final base diameter is achieved. After circulating at low velocity to clean the slurry, a sonic contour mapping tool (Mesotech or equal) is employed. This gives the contours at each horizon from top to bottom, at increments of 0.5 m. a reinforcing steel core, complex with spiral, is lowered to the bottom. It is spaced by pads to ensure that it is centered and has adequate cover outside the core. A steel tubular insert with closed end may be used instead.

Tremie concrete is then placed; typically the drill string is used as a tremie but a separate tremie pipe may also be employed. This concrete must be placed by tremie and pumping down must be prohibited. It, of course, is permissible and desirable to deliver the concrete by pump or conveyor to the open hopper at the top. Direct pumping to the bottom of the shaft has resulted in many failures. The exact causes have not been agreed but this author believes that the following explanation is valid. The pump periodically fills the entire pipe so that the entire head of fresh concrete acts at the discharge. Add to this the surges in pump pressure. The result is a high flow rate and a violent mixing action which washes the cement from the concrete mix.

The question has then been asked: why does grouting through a 20 to 30 mm diameter pipe not cause similar problems? The answer is that friction prevents high flow rates.

To return to the tremie concrete, it should follow general rules for the mass tremie concrete, but because of the typical smaller diameter of the tremie pipe, the coarse aggregate is usually limited to 10 mm maximum. An oversanded mix with sand constituting 50 per cent of the total aggregate should be used.

A belled shaft has to take high shear and deep beam bending. Yet no practical system has been found for placing reinforcement in the bell, outside the core. Therefore, until now, efforts have been concentrated on use of a low heat mix to prevent thermal cracking. The use of small-size aggregate and silica fume enhances the tensile strength. However, for the future, the inclusion of steel fibers in the tremie mix, as developed in the reconstruction of the Potsdammer Platz in Berlin, appears to offer a more positive solution.

As the concrete is being placed in the drilled shaft, common practice on the land has been to withdraw the casing. This same practice has been occasionally employed for overwater bridges but is not recommended. The withdrawal of the casing too often results in damage to the fresh concrete of the drilled shafts, such as slurry inclusions and voids. On the famous cable-stayed bridge across the mouth of the Seine in Normandy, the drilled shafts of a main pylon were found during construction to be defective because of voids. This resulted in a 1 year delay and a major claim.

This authority recommends leaving the steel tubular shell (casing) in place as a permanent member, working in composite action with the concrete and serving as bending reinforcement. The casing also protects the concrete and stiffens the pile. In effect, the concept merges with that of a concrete filled tubular pile.

A compromise solution has been developed in Hong Kong. After the shaft has been drilled, an interior corrugated shell of 50 mm less diameter is placed. The annulus between the shell and the casing is filled with pea gravel (rounded stones of uniform 8 to 10 mm size). Then the reinforcing core is set and the shaft is concreted. After the concrete has hardened, the outer casing is extracted by vibration. This method appears to give satisfactory results. However, due to the many failures of drilled shafts on land, requiring corrective action, it is both sound technically and economically to leave the casing in place.

The Submerged Prefabricated Tubes

More than 100 submerged prefabricated tunnels have been successfully constructed to cross estuaries, harbors, and rivers in the United States, Europe, Asia, and Australia. As a result of this repeated use, major improvements in construction and installation have been and are still being made.

These tube segments are large, typically 100 m long, 30 m wide, and 10 m high. They are prefabricated either as steel-concrete composite segments, favored in the United States, or as reinforced and prestressed concrete, favored in Europe. Each segment is then floated to the site, where a trench has been dredged and a stone base placed and screeded. The tube segment is supported from the surface by a catamaran barge and ballasted down onto the stone base or leveling pads. Connection is made to the previous segment. Finally, the tubes are backfilled and a protective cover of fill or riprap placed over the top. Regardless of which type of construction is used, the walls will be thick concrete, to give adequate weight on bottom when finally installed and dewatered.

The large size and mass of these structures requires careful consideration of the effect of currents, water density, as well as hydrodynamic and inertial forces while submerging through the several strata of water.

The steel structure, consisting of the external skin plate and the internal stiffening and bracing, is typically fabricated in a shipyard. Heavy temporary bulkheads are attached to both ends. These are designed to resists the maximum hydrostatic pressure during submergence. Gaskets are fitted. Reinforcing steel studs are attached to the skin plate and other studs to the bracing. The additional reinforcing steel for the concrete may also be installed at this time, as may the internal forms. The steel hull is then launched sideways, using standard slipways, and then moored at an outfitting dock. Working through holes in the roof, workers progressively construct the reinforced concrete. Tunnel-type forms are used, placing a 16 to 20 m long segment at a time. Concrete is delivered by pumping. Careful control is kept of draft. When concreting is finished, the freeboard will be about 1 m. A light structural tower and standpipe is installed on top of the tube at the loading end, to serve as a guide during installation and as access for the umbilicals carrying ballast lines, compressed air lines, electrical services, and instrumentation. The standpipe also enables worker access.

This same structure may be fabricated and completed in a gravity dock, as was done for the Third Boston Harbor Tube.

Prefabrication of All-Concrete Tube Segments

These are typically constructed in a graving docks or construction basin, so that all construction may be completed before float out. The reinforced concrete structure is constructed in short segments, using a concrete mix especially designed for low heat of hydration. Thermal effects generally control the design of the reinforcement to limit cracking.

In many European projects, the structure has been heavily prestressed transversely. Because of the thick sections, a large number of tendons are required to keep a significant compression stress across the section. Consideration must be given in the prestress design to the different stages through which the structure passes. During fabrication and prestressing, there is no hydrostatic load. At float-out there is hydrostatic load on the bottom and sides but none on the roof. In the final case, the roof is also loaded.

Careful control is kept on all dimensions during construction, since these determine the weight and displacement during float-out. Heavy temporary steel bulkheads are fitted to each end and gaskets placed. A tower and standpipe is attached, in order to provide for control and access. Waterproofing membranes are applied to the wall and roof. In the case of the roof, the membrane may be steel plate, attached by stud bolts, with grout injected underneath. A rubber membrane has also been used as the top membrane. The structure is floated in the dock and any needed adjustments in trim or heel are made by solid ballast.

Not all concrete tube has been prestressed. The most recent tube segments in Hong Kong, Sydney, and Denmark have been constructed with reinforcing steel only. The reinforcing steel has been detailed to limit any cracking to very small widths. Special care has been taken to limit the heat of hydration of the concrete. A waterproofing coat of plastic has been placed on the side and roof.

In the case of the Øresund Tunnel between Denmark and Sweden, a unique form of fabrication is being applied. Each large segment is subdivided into 5 m long segments of full cross section. These are constructed above sea level, in a factory. Each small transverse segment is match-cast against the preceding segment. Each small segment is cast in one continuous placement to eliminate the differential thermal contraction and shrinkage which causes cracks. As the small segments are completed, they are progressively jacked out on rails. Watershops are installed between each small segment. When all of the small segments for a 140-m-long segment are completed, they are temporarily post-tensioned together. Temporary end bulkheads are affixed as before. Using level and a temporary bulkhead gate, the water level is raised around the large segment until it floats. The segment is then moved seaward to a deeper basin, where the water level is lowered to that of the adjacent sea. Now the segment can be floated out, towed to the site, and submerged onto a pre-level bed. When the segments are seated and backfilled, the temporary post-tensioning is removed.

While the segments are being prefabricated, the trench is being excavated by dredging. It is then partially backfilled with crushed stone or gravel to bring the bottom to grade.

In the earlier tube projects, especially those in Europe, no attempt was made at exact grade but the stone or sand bag was leveled sufficiently by drag screeds to ensure that no place was above grade. This method requires that temporary seating pads be constructed to exact grade upon which to seat the tube segment.

In the case of the Webster Street Tube between Oakland and Alameda, California, the temporary pad was pile-supported with a special cushioning cross girder on which the tube segment temporarily rested. This cross girder was fabricated of timber and steel, to be strong enough to support the tube during initial seating but to crush when the tube had been backfilled and fully ballasted, to transfer the load uniformly to the sand. In another instance, polyurethane pads on top of concrete slabs were used, designed to transfer the load in similar fashion. In the case of several Danish-designed tubes, concrete pads were set, on which jacks operated from within the segment adjusted the segment to exact grade. Then sand was flowed underneath.

On most recent U.S. tubes, the stone bed screeded to exact grade. A special semisubmersible catamaran barge is employed, with pull-down devices at each corner, enabling the screeding barge to be kept at constant elevation despite the tidal changes. The pull-downs react against preset concrete weights. From the catamaran, longitudinal girders are hung to exact grade and profile. Running on these girder rails is a transverse hopper and screed. The hopper is fed by tremie pipes from surface. Typical tolerance achieved is ±20 mm.

On the Øresund Tunnel, the stone bed has been similarly screeded with a dredge ladder, fitted with a special head that distributes the stone. The head is computer controlled to exact elevation by electronic position and depth indicators. The dredge ladder and head sweep back and forth on a semicircular pattern to cover the entire area.

In most recent construction, the segments, 80 to 140 m long, are installed by means of a catamaran barge. This barge consists of two hulls which straddle the tube segment. The catamaran is moored in position with taut moors and the segment floated in and made secure. However, in the case of the Øresund Tunnel, pairs of pontoon barges, joined by a gantry truss overhead, in effect, two mini-catamarans, are being used to eliminate the longitudinal bending moments in the hull.

Stability during immersion is assured by keeping the center of gravity of the tube below the center of buoyancy, since the contribution to stability of the water plane is lost as the top passes below water. The ballast for overcoming the positive buoyancy which the tube had during float-out is usually placed as dry ballast, *e.g.* gravel, or is placed in compartments, so as not to develop a free surface which could reduce stability. In the case of the Webster Street Tube in Oakland, California, water was sprinkled on sand beds, to add weight with no free surface effects.

The tube segment, as it is lowered, is both affected by the current and, in turn, affects the current by causing it to speed up under the tube as it nears the bottom. This can erode the trench sides and even the screeded base. The tube segment should therefore be set during a period of low current, *e.g.* slack tide. Horizontal positioning is controlled by mooring lines, run from platforms on deck or from the shore or dolphins to the segments itself, so as not to impose any rotational movement on the tube segment. Water density may increase as the structure nears seafloor due to salinity and suspended silt. Cases have occurred where the tube segment would sink no further until additional ballasted was added.

A new tube segment is typically lowered 1 to 2 m seaward of the end of the previously placed segment. When it is 1 m above final landing, lines from fairleads on the previous segment pull the new segment to a 0.5 m gap. This gap may be surveyed by direct measurement by divers or by short-range sonic transducers on the tube end. The tube segment is then lowered to rest on two steps protruding from the previous segment. At the same time, the leading edge of the tube is kept 0.5 m above the stone bed or leveling pads at that end.

Come-along jacks, similar to railroad car couplers, are attached at each side of the previous tube and used to pull the new tube tight, compressing the soft portion of the GINA gasket. The leading edge is then set down on the leveling pads. Working from inside the previous tube, the space between the two closure bulkheads is drained. The full hydrostatic pressure acting on the leading end bulkhead is now available to squeeze the gasket tightly together. Additional ballast can now be added. When full seating and complete drainage has been verified, the two adjoining end bulkheads are cut out and the joint made good. For steel-composite tubes, this is accomplished by welding plates and grouting. For the reinforced concrete tubes, an inner Omega-type gasket is added.

When the tube segments have been seated on the steps and leveling pads, just above the stone bed, sand is flowed or jetted to fill the gap. Care has to be taken to avoid too great a hydraulic head of fluidized sand, so as not to raise the tube if the exist become blocked. Weak grout has similarly been injected under tube segments, the edges being sealed by sandbags or backfill. Initial backfilling around the sides of the tube must be done with care and consideration of symmetry to avoid sideways displacement of the tube. For that reason, placement through a large tremie tube is used rather than bottom dumping. Sand so placed will be of low density. In seismic regions, densification may be required by vibration or stone columns.

A sand blanket is then placed over the top of the tube. On this rock riprap or an articulated concrete mat is placed, in order to prevent damage to the tube from a dropped anchor. The riprap should be placed by skip or bucket and lowered to the seafloor so as not to damage the tube.

The ends of the tubes are connected to portal buildings in which the ventilation fans are housed. These are constructed separately, usually in a cofferdam of steel sheet piles. In the case of the Third Boston Harbor Tunnel, a circular concrete slurry wall cofferdam was used.

Joint details are similar but more complex than those between segments, since they must accommodate thermal and seismic expansion and contraction as well as differential settlement.

These have been proposed for many years for deep-water crossings and have been engineered in considerable detail for the crossings of the Strait of Messina and Hogsfjord, in Norway. Proposed construction procedures are similar to those for standard submerged prefabricated tunnels except that each tube segment must be held in position while tethers are attached.

The tethers run to anchor blocks, either gravity blocks placed on the seafloor or piled anchors. Because of the depth, the tethers will usually be pre-attached and fabricated with upper fittings to exact length. In the case of the piled anchor, a seafloor template can be secured to the piles at the exact design elevation. In the case of gravity anchor block, the settlement in the soil may be somewhat indeterminate so instead of pre-attaching tethers, they may be fabricated after the block location is known. Then the tethers can run down pre-attached guidelines to look into the block.

Tolerance and weight require closer monitoring than with the bottom-founded tubes because the tubes are in state of positive buoyancy; yet this must not overstress the tethers.

Various alternative solutions to the installation have been proposed, including towing the entire tube, with all segments preconnected on the surface. This appears to pose severe risks.

The Types of Barriers

Venice Storm Surge Barrier

This mammoth project, involving the closure of the three openings to the Venice Lagoon, has been quite thoroughly planned and engineered, but despite the need, not yet implemented. See Figure 11.2.

Designed with optimum hydraulic shape for the sill, and carrying the heavy steel gates, these highly irregular structures presented a serious problem of stability during submergence. Use of temporary vertical auxiliary buoyancy tanks appears practicable. Control of position at the site, with substantial current running, had been originally planned to be by means of lines from the towers. Studies showed, however, that this would exert tilting forces on the segments, so attachment at the center of rotation was selected instead.

The opening is to be first bounded by steel sheet pile walls at either side, then dredged to 1 m below founding grade. The heterogeneous lenses of clay and silty sand will be pinned together and densified by the driving of closely spaced reinforced concrete pile, driven with their heads terminating 1 m below founding elevation. A 1-m thick blanket of crushed rock is then placed.

The concrete still segments, with gates attached, will be ballasted down to found, very much in the manner described for the submerged

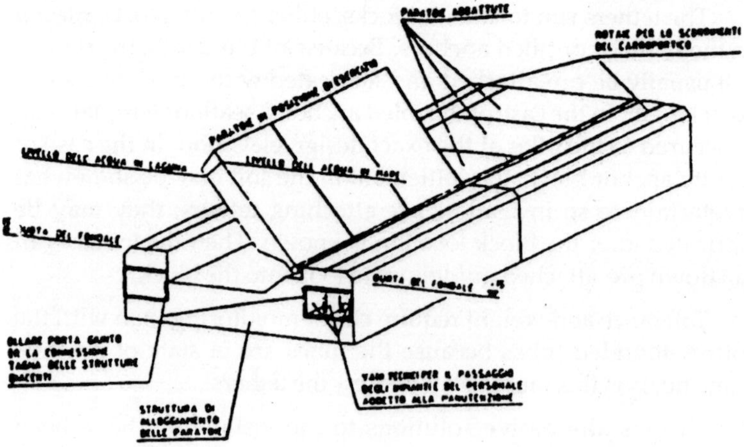

Figure 11.2: Venice Storm Surge Barrier Concept

prefabricated tunnel units. See Figure 11.3. Like them, each end will be seated on leveling pads of precast concrete, accurately preset to exact grade. Then the new segment will be slid sidewise, while it still has little net weight and hence little friction, squeezing the GINA gaskets between segments. Subsequent dewatering of the space between the two end bulkheads will make the fully hydrostatic head on the leading end available to compact the gaskets. See Figure 11.4. A steel sheet pile wall will then be driven on the seaward face to prevent piping and flow under the sill.

This project has unfortunately not yet been implemented despite the obvious growing need to prevent repeated flooding of Venice.

The barriers represent a hybrid in construction technology between bridge piers and submerged prefabricated tunnels. They are prefabricated, typically in a construction basin, and floated to the site, sometimes partially supported by external buoyancy tanks or a catamaran. On arrival they are set down on a prepared base and joined to the previously set segment, muck like a tube. Gates are either pre-installed or are placed afterward, fitting in recesses.

Since they act as a low-head dam, underflow must be prevented. In most cases this is accomplished by the installation of driven steel sheet piles, although in the case of the Oosterschelde Storm Surge Barrier, the Dutch elected to lengthen the path of permeation

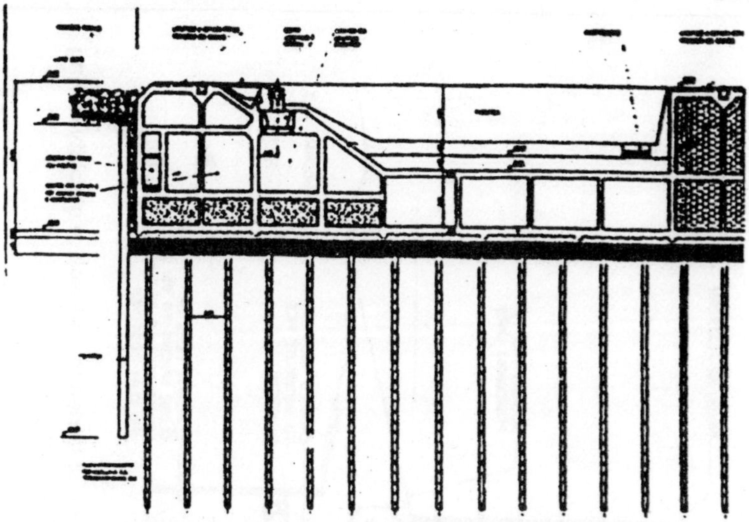

Figure 11.3: Transverse Cross Section of Venice Storm Surge Barrier

horizontally so that underflow pressures would be fully dissipated. They placed filter fabric and mattresses over extensive areas each side of the barrier. This meant that the permeation would be limited even if settlements occurred.

Oosterschelde Storm Surge Barrier

The Oosterschelde Storm Surge Barrier, across the mouth of the Eastern Scheldt Estuary in The Netherlands, must rank as one of the major offshore engineering and construction achievements of the century. Although constructed in water depths only 20 to 40 m deep, it is exposed to waves and winds from the North Sea and to high tide currents, which require the development of new techniques for offshore construction which will have application on future projects. The project is also notable because of the innovative development of specialized construction equipment.

In the project, 66 mammoth concrete gate piers are being installed, seated on a prepared foundation on the sands of the river delta. Initial operations offshore commenced with construction of an island near the center of the project from which all construction

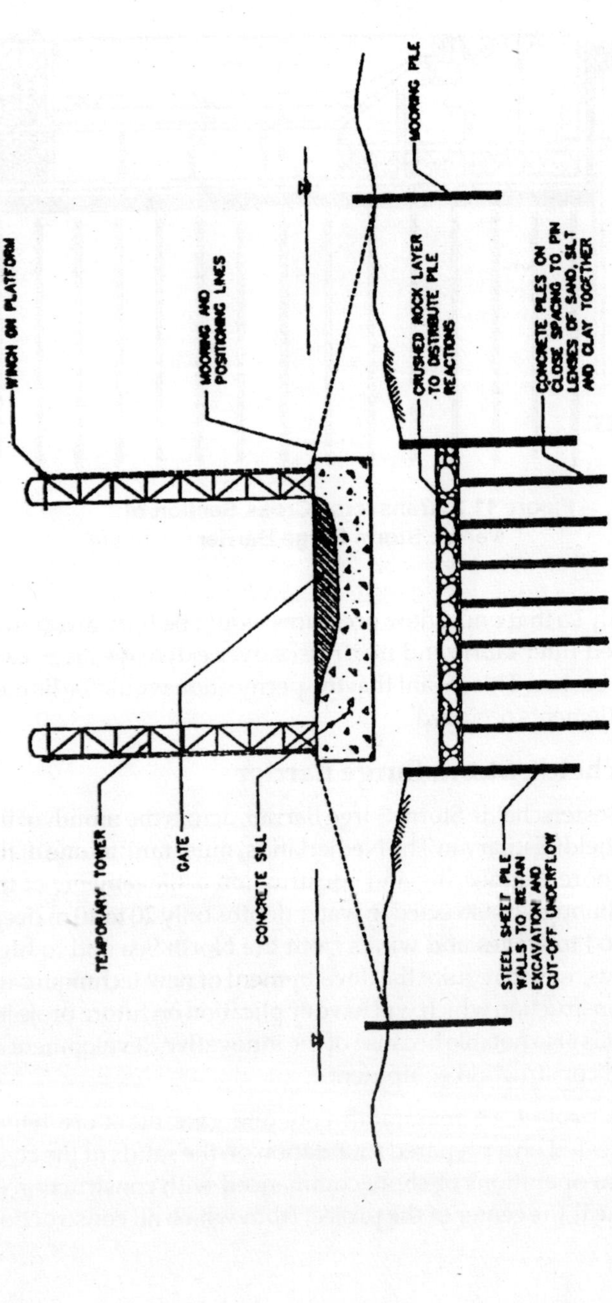

Figure 11.4: Conceptual Installation Procedure. Note erroneous location of fairleads and lack of stability during immersion. Procedures were revised correctly.

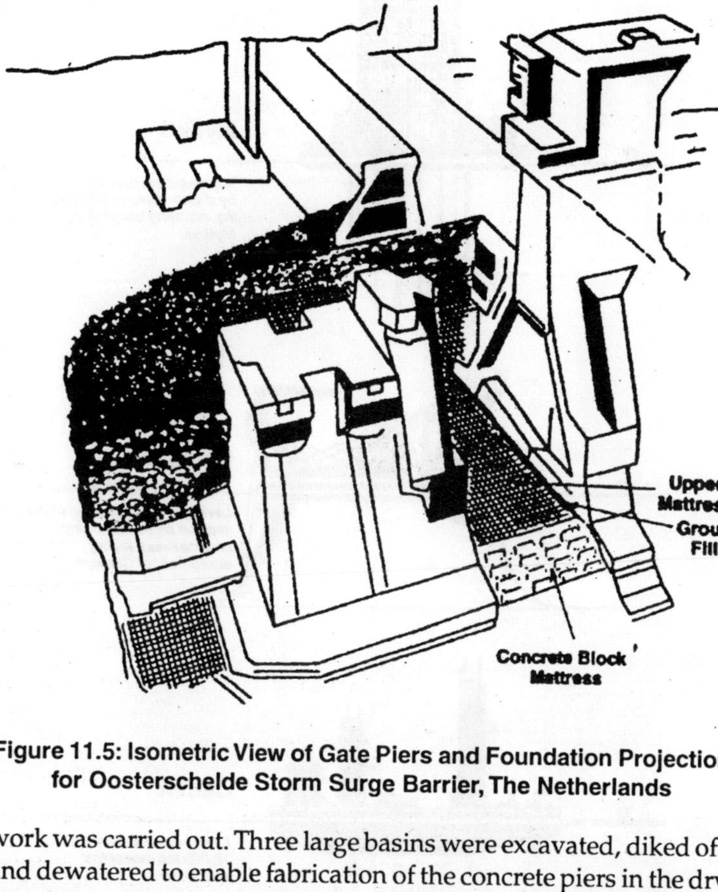

Figure 11.5: Isometric View of Gate Piers and Foundation Projection for Oosterschelde Storm Surge Barrier, The Netherlands

work was carried out. Three large basins were excavated, diked off, and dewatered to enable fabrication of the concrete piers in the dry. Simultaneously, the extensive slope projection work was carried out on the beaches and dikes adjoined the barrier proper. Aprons of asphalt-filled stone, sand-asphalt, and articulated concrete mats were laid.

Meanwhile, at the site, large-diameter dolphin and anchor piles (steel cylinder piles) were driven to serve as moorings for the extensive floating construction operations to come. Lines from the anchor piles were run up to mooring buoys. The loose sands in the top 10 to 20 m of the foundation under the barrier.

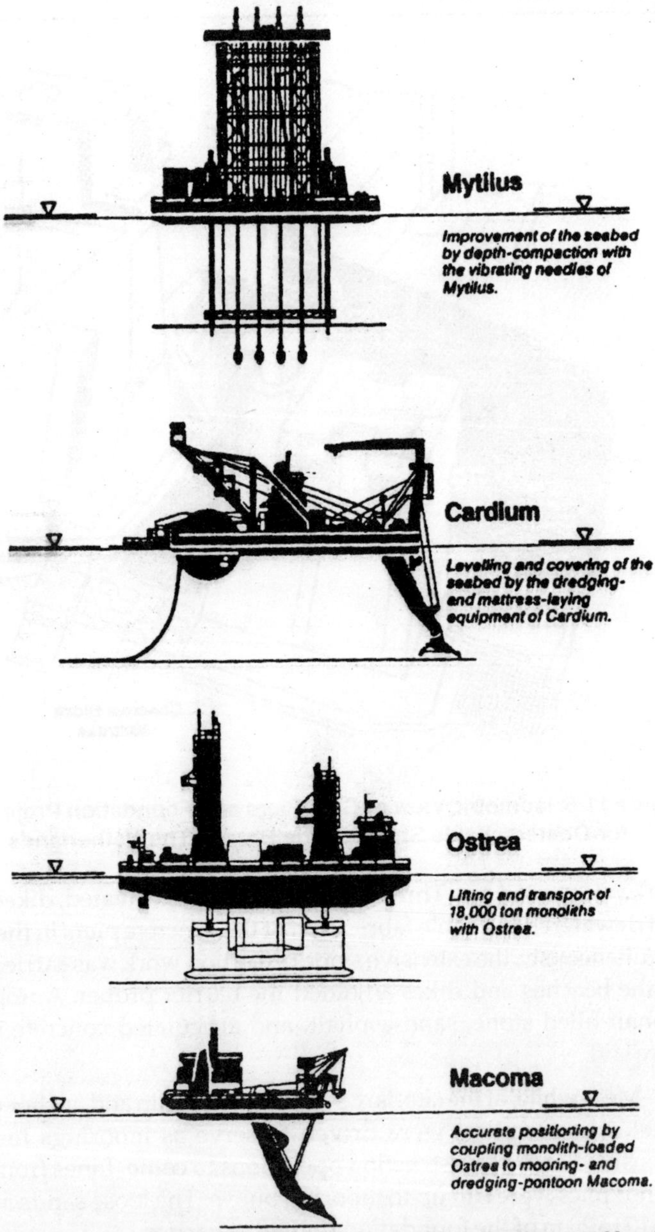

Figure 11.6: Specialized Offshore Construction Equipment Used to Construct the Oosterschelde Storm Surge Barrier, The Netherlands

Chapter 12

The Applications of Prime Movers, Generators, and Transformers

Introduction

Today the selection of prime movers and generators takes an important role in offshore engineering.

Now let us try to learn about Aero engine derivative set, and about gas turbines. There are two forms of gas turbine in use:

Aero Engine Derived

This consists of a modified aeronautical jet engine known as the gas generator which exhausts into a separate power turbine. This combination often produces a unit with very good power-to-weight ratio, as the gas generator is lighter than the integral unit on the equivalent industrial machine. However, the unit may require better protection from the environment and in some cases shorter intervals between servicing. Examples of this type of machine are Rolls-Royce Avon and RB211 based sets. Figure 12.1 illustrates a typical example.

Industrial

These are propose-built engines, which incorporate the gas generator and power turbine in a single design. The older machines

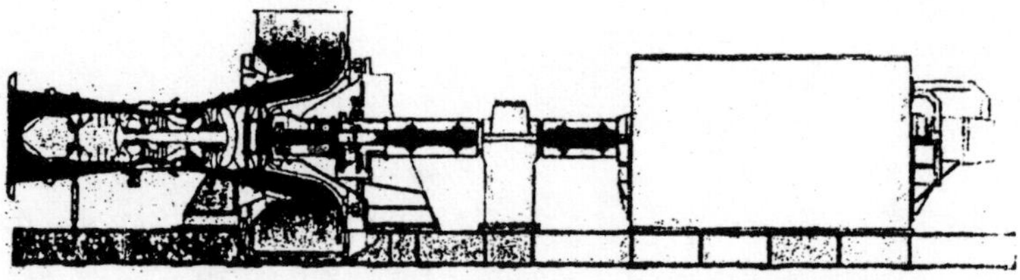

Figure 12.1: Aero Engine Derivative Generator Set.
(*a*) Rolls Royce Industrial Avon powered compression set being installed on the Brunei Shell Petroleum Company platform;
(b) sectional drawing of set shown in (a).

tended to be less fuel efficient than equivalent aero engine derived types, but they have a good reputation for reliability and for the toleration of fuel supply or load abnormalities. Examples of this type of machines are the General Electric and John Brown Frame 5 and the Ruston TB series. Figure 12.2 illustrates a typical example.

In most offshore situations (with the exception of storage and pumping stations) the gas is being produced via the process plant from a production well. Changes in well gas content, calorific values, pressure etc. can have significant effects on engine power output, which may in turn affect the production process and the fuel supply. Slugs of condensate may also appear in the fuel gas supply, having a similar effect. Although these unpredictable phenomena are avoided as much as possible by good process design, it is worth considering some means of catering for them such as one or more of the following remedies:

1. A small separator or knockout pot may be located close to the engine intake in order to remove condensate. If the supply pipe is long, it will be necessary to provide this anyway to remove condensate which has condensed in the pipe.
2. The engine combustion system may be modified for dual-fuel operation. When set up properly, this system can automatically transfer from gas to diesel combustion during fuel gas disturbances with negligible effect on engine power output.

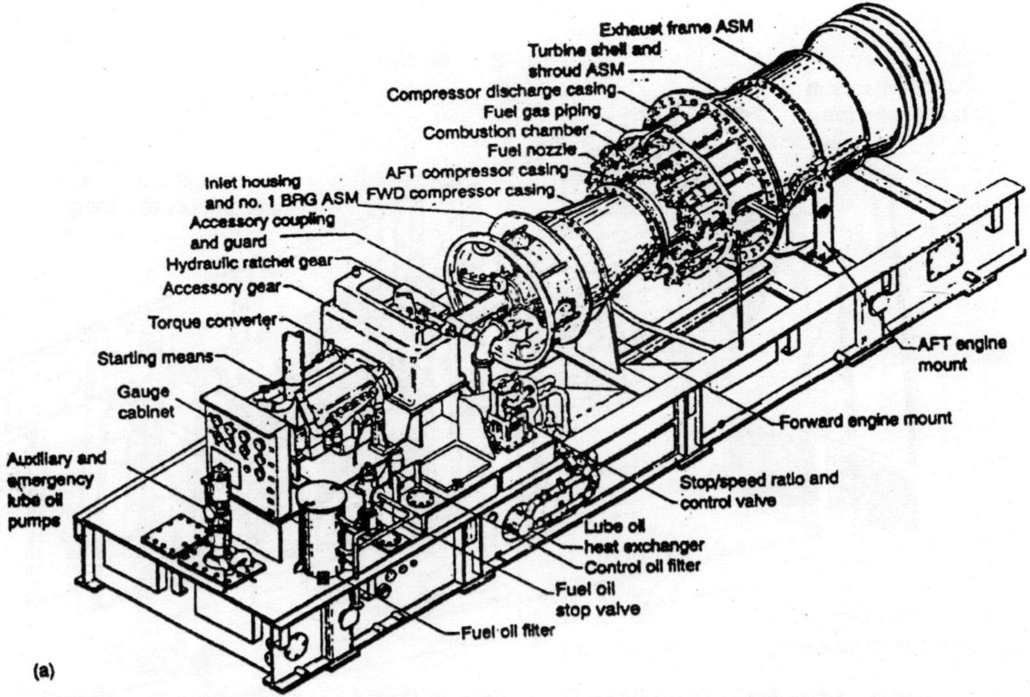

(a)

Figure 12.2: Industrial Engine Based Generator Set; (*a*) Engine, (*b*) Generator. (a) Courtesy of John Brown Engineering to GE (USA) design), (b) Courtesy of Hawker Siddeley Electrical Machines Ltd.)

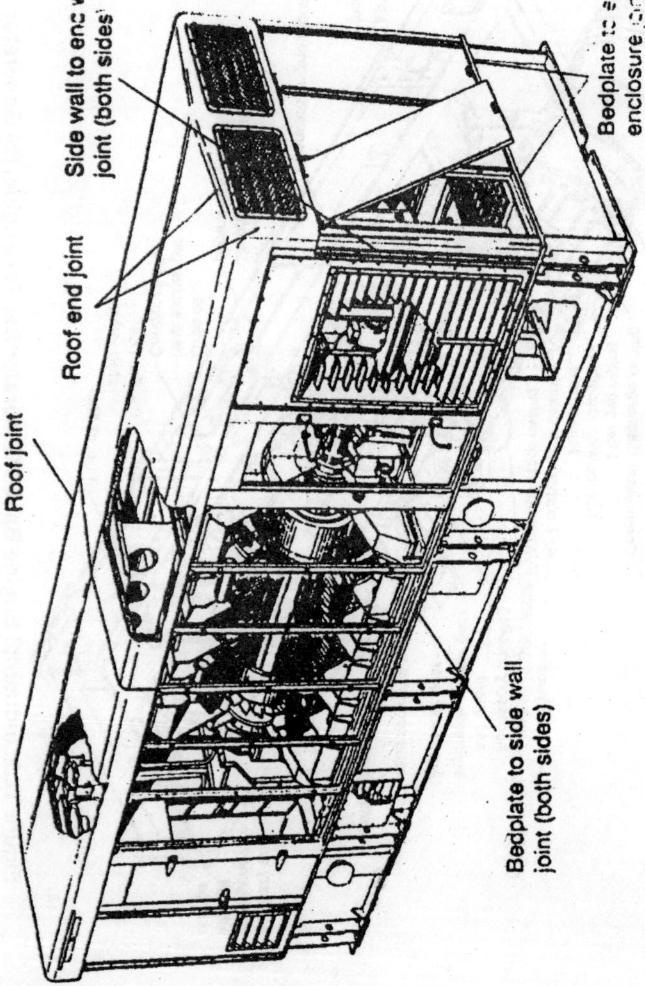
Roof joint
Roof end joint
Side wall to end wall
joint (both sides)
Bedplate to side wall
joint (both sides)
Bedplate to end of
enclosure

Figure 12.2: (b)

A large gas turbine generator set may take in the region of 30 seconds, from the initiation of an emergency stop signal, to slow down to a speed at which acceleration is no longer viable without cranking. During this period of deceleration, it is vulnerable to ingestion of gas from serious process gas leaks on platform. Such ingestion of gas in sufficient quantity, bearing in mind that all electrical loads would probably have been removed by the ESD systems, may cause the machine to reaccelerate and overspeed to destruction.

The risk of this occurring must be minimized. First, there should be careful sitting of gas detectors to ensure that serious gas leaks are detected as soon as possible. A 2-out-of-*N* voting system should be used to minimize spurious generator trips due to faulty detectors. Secondly, governor response to an emergency engine stop signal should be as fast as possible within the metallurgical limitations imposed by turbine blade cooling rates etc.

The Gas Turbine Generator Sets and Governors

On large gas turbine generator sets, lubrication is accomplished by a forced feed lube oil system, complete with tank, pumps, coolers, filters, and valves. Lubricating oil is circulated to the main bearings, flexible couplings and gearboxes. A portion of the oil may be diverted to function as hydraulic oil for operation of guide vanes etc. within the turbine. A typical arrangement is for the lube oil pumps to take their suction from the lube oil tank and the hydraulic control valves to take their suction from a bearing header. The system may contain between 2000 and 10,000 litres of oil.

The electrical engineer's interest in this system is that out of, say, six lube oil pumps on each generator set, five are driven by electric motors. The following are the description of the function of such pumps:

Main Pump

The main lube oil pump is a shaft driven positive displacement unit, mounted into the inboard wall of the lower casing of the accessory gear. It is driven by a splined quill shaft from the lower drive gear and the pressure is 65 psig maximum. As the output pressure of this pump is engine speed dependent, with certain models of turbine insufficient lubrication pressure is available from the pump below a certain speed. In this case, even when it is operating satisfactorily, the pump may require to be supplemented by an electrically driven pump.

Auxiliaries Pump

The auxiliaries pump, mounted on the oil tank cover, is a submerged centrifugal pump which provides lube pressure during startup and shutdown of the generator under normal conditions. The auxiliaries pump is driven by a low-voltage 30 hp two-pole AC flameproof motor.

Emergency Pump

The emergency pump is also mounted on the tank cover, and is a submerged centrifugal type which also provides lube pressure under startup or shutdown conditions. This pump is driven by a 125 V 5 hp DC flameproof motor.

The three remaining pumps of our typical system are:

Main Hydraulic Supply Pump, Auxiliary Hydraulic Pump

Failure of these pumps and the resulting low hydraulic pressure would not necessarily cause an immediate generator set failure, although the unit would eventually trip owing to low lube pressure.

Hydraulic Ratchet Pump

Loss of hydraulic ratcheting pressure or equipment would cause generator shaft bowing and would also lead to excessive bearing stress. It may, however, be possible to repair the fault by, for example, replacing the motor within few hours, *i.e.* before serious damage has been caused.

These three pumps would normally all be driven by low-voltage AC motors.

The basic criteria for establishing electrical supplies to generator lube oil pumps and other vital auxiliaries are as follows:

Although weight and space limitations usually prevent auxiliary switchboards being fed from separate generator transformers, as one would expect to find in a power station onshore, the switchboard or individual motor starters should be electrically as close to the generator as possible.

Each complete set of electrically driven auxiliaries should be supplied from one switchboard, so that only one supply is required to ensure that the particular set of auxiliaries is available.

Any standby auxiliaries should be fed from another switchboard, which obtains its supply from the generator via a different electrical route.

If possible, the complete set of auxiliaries should be duplicated in order that a fault on one switchboard or its incoming supply will not lead to a shutdown of the generator.

Loss or temporary disconnection of supplies to the emergency DC lube oil pumps should be adequately displayed on the generator control panel annunciator, so that the operator is aware that loss of AC power, and consequently the failure of the AC driven lube oil pumps, may lead to damage to the generator as it runs down.

Now let us try to understand about electric governors.

Sophisticated electronic governors able to provide reliable service coupled with high performance have been available for some time. However, there are a few points worth considering when selecting a suitable governor. First, it is advisable to obtain as many of the governor's control system parameters as possible for the governor manufacturer. These will be required if the power system is to be computer simulated before the project design stage is complete, which is usually the case.

Any governor-engine-alternator system will have a finite response time. It also has a certain amount of stored energy which can be extracted to maintain the system by sacrificing some speed, and this energy may be used while the governor is responding to a load increase. Apart from large power system disturbances such as large motors starting or short-circuit faults, which need to be studied by dynamic simulation, any cyclic loads need to be considered carefully. The following anecdote illustrates the problem.

A radio transmitting station was built in an isolated location without access to any external electricity supply, depending entirely on two diesel generators for electrical power. The station was successfully commissioned, with the exception that the supply frequency became unstable and the generators shut down during each broadcast of the time pips. The problem was traced to the cycle loading imposed by the series of pips being at as frequency at or close to the natural frequency of the governor mechanical linkages. In this case the problem was overcome by substantially increasing the mass of the engine flywheel. The only cyclic loads likely to be experienced offshore are those produced by large reciprocating pumps and compressors or by heaters which use thyristor integral cycle firing controllers. The phase angle controlled rectifiers and variable frequency inverters used on the drillings rigs are unlikely to produce the necessary supply frequency subharmonics.

The control system parameters for the generator excitation system, like those for the governor, will be required if any computer simulation work is to be carried out.

The following aspects need particular attention in offshore systems:

Although, as with large onshore machines, two-pole 50 or 60 Hz designs are used because of the greater efficiency of energy transfer at the higher speeds, it should be remembered that this requires complex engineering analysis and the use of high-grade materials, particularly as the generator module could be located on a 200 m high steel structure, 30 m above sea level.

To the author's knowledge, the highest generator rated voltage offshore is 13.8 kV, and alternator manufactures have no difficulty in producing machines at this rating. Voltages up to 22kV could be used, however, provided that suitable switchgear is available.

The alternator subtransient reactance X"d is a useful regulator of maximum prospective fault current, and alternator manufactures are usually prepared, within certain limits, to vary the design of the windings to enable the system designers to limit the prospective fault level to a value suitable for the switchgear available. For a machine of, say, 30 MW, the degree of variation for X"d would be approximately 15 per cent to 21 per cent.

Although, as discussed above, a highly reactive machine may be beneficial in order to limit prospective fault currents, this is accompanied by a penalty of poor motor starting performance due to the increased transient reactance (X"d). The winding reactance must therefore be optimized for the best motor starting performance, allied with prospective fault capabilities within the capacity of the switchgear installed. This tradeoff is best accomplished using computer simulation.

The conventional configuration of brushless alternator with pilot and main exciter is commonly used offshore for machines of 500 kW rating and above. Machines with static (rectifier derived) excitation are acceptable provided fault currents can be maintained for at least the full generator fault time rating. A definite time overcurrent device should be used to shut down the generator within this time, so that the machine is not left running with a fault still on the system after the voltage has collapsed.

The automatic voltage regulator should incorporate a means of detecting a control loop disconnection such as that caused by open-circuit voltage transformer fuses, in order to avoid excessive voltages being developed on the machine stator if such a disconnection occurs.

Most offshore production installations have three or four main levels of operation which are reflected in control systems such as the ESD system. If, however, there is a very large gas leak such that the installation is enveloped in a gas cloud, it would be necessary to isolate all forms of electrical power capable of igniting the gas, including in some instance the DC secure supply batteries.

The next level of operation is with the emergency generator only running. The provision of an emergency generator is a statutory safety requirement, and as such it should be designed to provide reliable power for statutory communications equipment, navigational aids, fire and gas monitoring, ballast systems and (although not statutory requirements) accommodation cooking, drinking water and sanitation facilities. As this generator must not be dependent on the platform production processes for fuel, it is invariably diesel driven. Storage of petrol or propane on the platform would be considered a hazard, which would rule out the use of an ignition engine for this purpose. These generators are usually designed to be automatically started on failure of other, larger generators on the installation. Again, there is a statutory requirement that the starting equipment for this generator is capable of at least six start attempts.

This generator should be located in a safe area, close to the accommodation, radio room and process control room. A day tank is required near the generator, big enough to run the machine for the time specified in the relevant statutory regulation. The time will vary depending on other installation conditions, such as whether it is regarded as manned or unmanned, but may be 24, 48 or even 96 hours.

The following points are often overlooked in specifications for emergency generator sets:

Despite the small size of the prime mover, air intakes must still be provided with spark arresting devices and overspeed flap valves, and exhausts with spark arresters.

Interlocking facilities must be provided to ensure that the generator circuit breaker cannot close to an existing fault when the generator is automatically started.

Means should be provided to maintain the generator output current in the event of a fault, for long enough to operate protection devices, where this is possible with the limited magnitude of fault current available from such a machine. Leaving the machine running with collapsed excitation is dangerous, as the fault may disappear, to be followed by a sudden and possibly unexpected reappearance of full voltage on the system.

The Cartridge Fuses and Air Circuit Breaker

A fuse is basically a device with a central conductor that is designed to melt under fault conditions. Interruption is achieved by spacing the two ends sufficiently for apart for the arc to be naturally extinguished. However, to obtain a consistent performance characteristic, fuse elements are carefully designed for particular voltage and current ratings. The ceramic cartridge tube is filled with powdered quartz and sealed. The silver element is not a continuous strip of silver, but is necked in short sections to reduce pre-arcing time. It may also have sections of low melting point. M-effect material to improve performance at low fault levels (see Figure 12.3). The physical design of the element, such as the length and shape of the necks, will depend on the application of the fuse and its operating voltage. If the maximum are voltage is exceeded, the pre-arcing time will be reduced and the fuse will not operate according to its standard characteristic.

Since a fuse is purely a means of protection against overload and fault currents, isolators or contractors are installed in series to carry out normal switching operations.

The advantage in using fuses is that the resulting switch-fuse device is often less expensive, smaller and lighter than the equivalent circuit breaker, particularly in the higher power ranges. Exceptions to this for low-voltage distribution equipment are some of the higher-performance current limiting miniature circuit breakers. These will be discussed later in the chapter.

The advantages with fusegear are twofold. First, HRC fuses must be replaced once operated. No matter how reliable the circuit is, some holding of spare fuse will be necessary, and replacement of the larger bolted fuses is time consuming. Permanent power fuses are available, however, which use the thermodynamic characteristic of liquid sodium to interrupt fault currents for a few milliseconds, during which time a suitable switching device is operated to isolate the faulty circuit. This type of fuse is not used offshore. Secondly, the standard ranges of fuses available are limited in current and fault capacity below the maximum ratings available in switchgear ranges.

Now let us try to understand about Air circuit breakers and Bulk oil circuit breakers.

An air circuit breaker (ACB) is a device where the circuit is made or interrupted by moving contacts located in atmosphere air. The ACB relies on moving the contacts sufficient far apart to extinguish the are under short-circuit conditions. To assist in this operation, the design of ACBs has been greatly improved since the original invention, notably in the following ways:

'Trip-free' mechanisms are used, in which the geometry of the mechanism is such that tripping can occur even while a closing operation is under way.

Separate arc and current carrying contacts give better thermal rating.

Blowout coils and arc chutes are employed. A blowout coil uses the fault current to produce a strong magnetic field which pulls the arc away from the contacts into the arc chute. The arc chute is a series of parallel insulators designed to extinguish the arc by splitting and cooling it. Because of the need for blowout coils, performance will be partially related to the fault current magnitude. This effect will appear as a critical current below which the arc is not drawn into the chute; arc contact wear will be accelerated, and the risk that fault clearance will not be achieved will be increased.

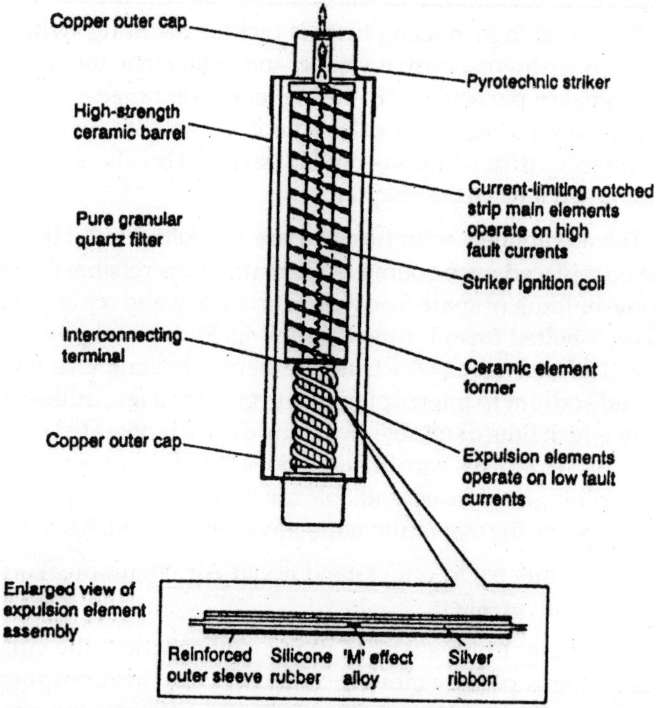

Figure 12.3: Diagram Showing the Interior of Typical HRC Fuse

This type of switchgear also tends to be bulky, mechanically complex and hence more costly compared with other forms now available. However, this equipment is well proven and its use well established offshore, since it has the ability to interrupt the high magnitudes of fault current found in offshore systems.

The bulk oil circuit breaker is a device in which the moving contacts are totally immersed in a container of mineral oil. As the oil is a better insulant and has a higher specific heat than air, contact gaps may be reduced, and the better heat dissipation means that the overall cubicle size may be reduced compared with those for an ACB of the same rating. Arc extinguishing is assisted by convection currents in the oil, produced by the heating effects of the arc. No weight saving is likely, however, because of the weight of oil required.

Arcing within the oil causes hydrogen gassing which must be vented for the container; although with modern designs the risk is minimal, open flames or even explosions may occur under severe fault clearing duty or normal fault clearing with badly contaminated oil. The insulating oil can deteriorate over a number of operations or after a long period, and requires regular sampling and testing. Sufficient replacement oil must be stocked offshore or at least made available offshore prior to testing.

Although this form of circuit breaker is still in common use onshore and has been installed for main switchboards offshore, it is now generally considered undesirable in an offshore environment and its use may be questioned by a number of certifying authorities and underwriters.

It is very much needed to study about vacuum interrupter which is the main part of vacuum circuit breaker.

As the heart of the vacuum circuit breaker is a device called the vacuum interrupter, one of which is required for each pole. The interrupter consists of a ceramic tube with metal seals at both ends. The fixed contact is mounted on one metal seal, while the moving contact at the other end of the tube is free to move inside a metal bellows which maintains the vacuum seal. The ceramic tube is usually in two parts, to allow the insertion of a sputter shield designed to prevent contact metal condensing on the ceramic tube and providing a conducting path between poles. On modern interrupters the sputter shield is now an integral part of the contact assembly, as can be seen in Figure 12.4. The vacuum is much harder, at 10^{-8} to 10^{-5}m bar, than in a fluorescent lamp, and should only allow a few free molecules. Thus the majority of ionized particles required to support an arc are provided by molecules from contact metal, the metallurgy of which is vital to the satisfactory operation of the interrupter. Hard contact metals will not provide sufficient molecules and arcing will be extinguished prematurely, leading to current chopping and high-voltage transient. However, if the contact metal is too soft, contact wear will be accelerated and contactor welding may occur. The difference in contactor metal is one of the essential differences between the interrupter and the contract bottle, the contractor having a softer contact metal. Vacuum contractors still have a significant fault rating, of around 7 kA, but, as with other forms of contractor, require to be protected by fuses for currents of

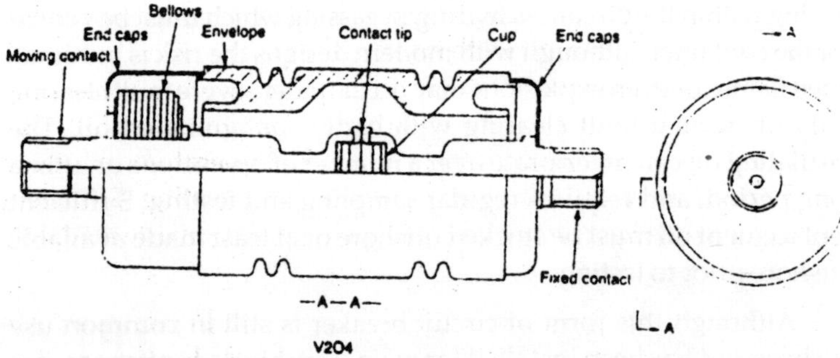

Figure 12.4: Interior of Typical Vacuum Interrupter

greater magnitude if contact welding is to be avoided. The circuit breaker resulting from this technology is compact and light, with a much reduced maintenance requirement. Foundation shock due to operation is very much reduced, owing to the lightness of contacts and the small distances they have to move apart for fault clearance.

Interrupters and contractors can be checked for adequate vacuum by applying a test voltage across the contact when in the open position.

A potential of 25 kV AC is required for testing a 12kV interrupter. Owing to the permeability of the various ceramic and metal construction materials, the operational life is limited because of a slow loss of vacuum, but by using a getter it is normally in excess of 20 years.

Short-circuit ratings if up to about 40kV are available for circuit breakers operating up to 12kV. This may not be sufficiently on some of the larger installations, where prospective fault currents and operating voltages may be higher.

Because of the size, weight and space advantages of vacuum interrupters and contactors, and the reduced fire hazard, this type of equipment is recommended for offshore use.

The Offshore Switchboards

In all cases, offshore switchboards must be highly resistant to the salt-laden corrosive atmosphere, and sufficiently moisture protecting to prevent the ingress of water, particularly from above (IP54 minimum). The enclosures must also physically protect operators from any arcing, flames or flying fragments due to maloperation.

Major changes to switchboards, once they are installed offshore, are extremely expensive. Depending on the planned operational life expectancy of the offshore installation, as much spare equipped capacity should be incorporated into the switchboard as possible. The spare auxiliary contacts of circuit breakers, contactors, switches and relays should all be wired out to terminals as a matter of course by the manufacturer. Whether racked or single mounting methods are used for protection relays, space for additional relays or spare relay rack spaces should be provided. Facilities for adding extra switchboard cubicles at either end should also be provided.

It is also important that facilities for cabling are as flexible as possible, and that sufficient space is available for installation of the largest cables without exceeding bending radii limits. Neutral connection arrangements for outing supplies with neutrals must be provided. Facilities need to be provided for the earthing of cable armour. If cable entry is at the top, there should be no risk of moisture entering via the cable entry even if the cables do not have drip loops.

The space requirements for circuit breaker handling trucks must not be forgotten.

Functional test facilities should be built into the switchboard using a test panel cubicle with bus wiring to each unit to avoid dangerous trailing leads. The test facilities must be interlocked to ensure that live operation of field equipment is impossible using the test panel. Every unit on the switchboard will require suitable permanent labelling and interlocking, shuttering and maintenance padlocking facilities. In fact, all the usual requirements for onshore substations such as safety rubber matting, earthing facilities and safety testing equipment will be required.

Now let us try to learn few things about emergency switchboard, and about drilling electrical system.

It is beneficial to provide synchronizing facilities for the switchboard's associated emergency generator. The generator has automatic start facilities which will initiate start following a main generation failure, provided the start signal is not inhibited by one of the safety systems. The synchronizing facility gives a convenient means of routine load testing for the generator, and allows for changing over to main generation after a shutdown incident, without a break in the supply.

The drilling electrical system is usually independent of the installation system, with its own diesel generation. The reason for this is partly to do with organization, since the drilling may be carried out by a different company which provides the complete drilling package, including generation and switchgear.

If the drilling system is of the same operating frequency as the rest of the platform, then an interconnector of some kind between the systems is mutually beneficial, there is no equipment in the main system that is particularly sensitive to the harmonies generated by the silicon controlled rectifier (SCR) equipment. The reliability and maintainability of the drilling electrical system is vital, as failures at particular times in the drilling operation may increase the risk of blowouts or cause the abandonment of the producing well.

The switchgear used for the drilling distribution is of the conventional motor control centre type, with the exception of the SCR cubicles. The SCR cubicles cannot be isolated individually owing to the permanent interconnection arrangement between each of the variable speed drives. An assignment switch on the driller's console allows the connection of any SCR cubicle to any DC drive motor in the system.

This arrangement allows for DC drive motors to be reassigned to another SCR cubicle if a fault develops in the first cubicle. To allow plenty of ventilation, the SCR cubicles are of a much more open design than other offshore switchboards. The assignment contactors are usually arranged in the upper section of the panel, while the SCR assemblies are in the lower section.

To avoid obstruction of cooling air, there are usually no insulating barriers between the interconnecting busbars, contractors, and equipment within the cubicles. However, it is sometimes necessary, if drilling is to continue, for the rig electrician to change SCRs with the cubicle only isolated by the assignment contractor at the top. This problem is usually overcome by using a removable insulating barrier which can be carefully located below the contactor before working on the SCR assembly.

Electrical supplies for living accommodation are important for the well-being of the offshore staff. Failure will not only bring

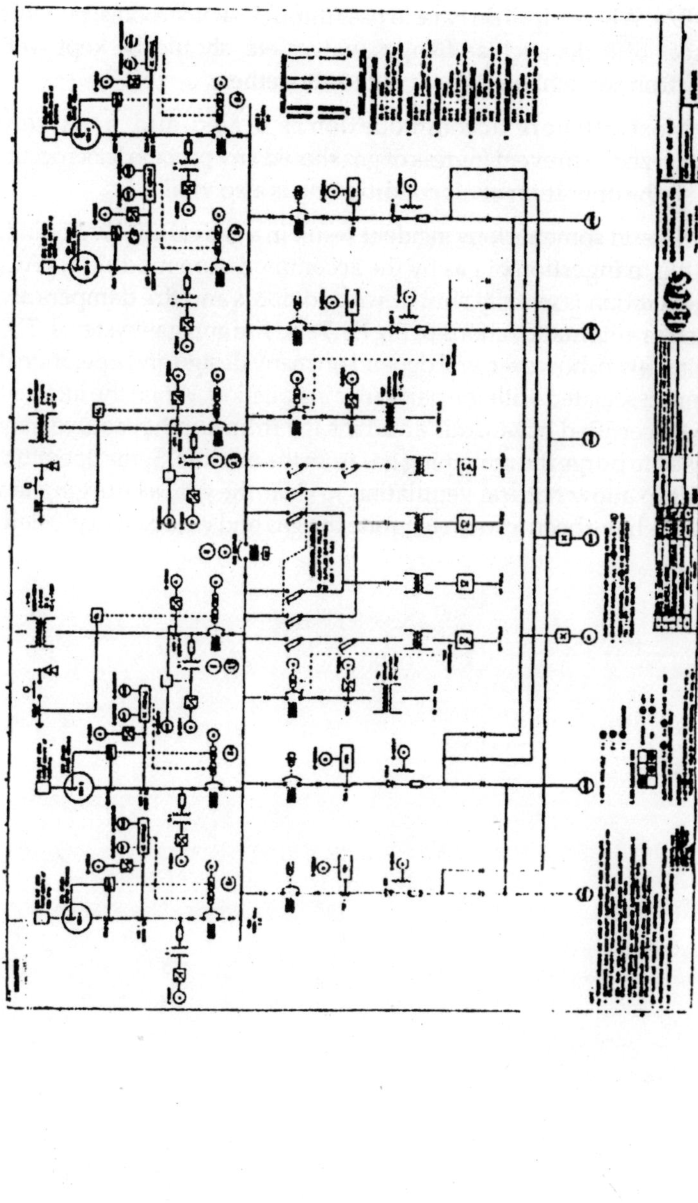

Figure 12.5: Schematic Diagram of Typical Drilling SCR Cubicle

discomfort, but will present a serious health hazard, when frozen food begins to thaw, toilets cannot be flushed etc.

Domestic pipework should be kept out of switchrooms where possible. Where pipes do have to pass through switchrooms, possible sources of leaks such as flanges, valves etc. should be kept well away from switchboards or avoided altogether.

Most offshore accommodation is sealed and positively pressurized to prevent ingress of gas should any process leaks occur, and so the operation of air conditioning is also vital.

Should some serious incident result in a gas cloud developing, leading to ingestion of gas by the accommodation module HVAC, all generation is usually shut down and doors and fire dampers are closed as automatic actions of the ESD and fire and gas systems. The actions taken, however, will depend on many design and operational factors associated with the particular installation. When the incident is over, certified hazardous area fans fed from another source may be used to purge the remaining gas from the module. Some operators prefer to allow natural ventilation to clear the gas, as purging fan systems have been known to ignite the gas and cause an explosion.

Chapter 13

Protection of Electrical Equipments, Used in Offshore Construction

Introduction

The isolated location, onerous operating conditions and harsh environment offshore make it particularly important that provisions are made in the design in order to protect electrical equipment from faults in the system, and to remove faulty equipment from the system quickly and safely. Alternators must be protected from prime mover and fuel system faults. Motors must be protected from faults in driven equipment.

Medium-voltage generators are normally resistance earthed in order to limit fault currents. Stator winding should be designed to minimize third-harmonic circulating currents, so that generators may be paralleled without extra weight and space being taken up with earthing circuit breakers. The short time rating of any earthing resistors must be borne in mind when calculating earth fault relay settings. A current sensing relay of either inverse or definite time characteristic may be used for unrestricted earth fault protection. Alternatively, an attracted armature relay may be used in conjunction with a time delay relay. In either case, the time delay must be within the time rating of the earthing resistor and the relay must coordinate with those downstream. The term 'backup' earth fault protection is sometimes used to describe this scheme, as it supplements the generator differential protection. However, unrestricted earth fault protection will not detect phase-to-phase faults, and therefore the term 'backup' or 'standby' leads to some confusion and is not recommended.

The neutral earthing conductor arrangement must be rated for the prospective earth fault current flows available from the generator. If the generator is a low-voltage machine with bolted earth connection, this current is likely to be high enough to affect cable conductor sizes. With directly earthed low-voltage machines, the high currents are used to operate fuses and miniature circuit breakers (MCBs) usually found in low-voltage distribution systems.

With larger medium-voltage machines the damage caused to laminations is limited by the insertion of the neutral resistance, and more sensitive forms or protection may be used in the medium-voltage distribution system.

The large offshore alternators will not have overload protection as such, but resistance temperature devices (RTDs) are buried in the stator winding, and will give alarm and trip signals at temperatures where overloading or other abnormal conditions such as excitation faults have occurred. The temperature protection will avoid unnecessary shortening of insulation life.

With small emergency generators, the associated circuit breaker should be provided with a thermal element.

To protect the generator from downstream faults, overcurrent relays of either the inverse definite minimum time (IDMT) or the definite time type are required.

The protection relay must provide sufficient time for downstream protection relays to operate to clear the fault, but this time must be within the time rating of the alternator. Large generators are usually provided with permanent magnet pilot exciters to maintain a fault current of around three times rated current continuously after the transient short-circuit current has decayed away. This can be maintained usually for an absolute maximum of 10 seconds before the generator overheats.

However, it should not be necessary to approach even 5 seconds in order to operate downstream protection, and the damage due to arcing fault if such a large amount of fault energy is let through can be extensive. It is therefore advisable to keep fault times to a minimum by good relay coordination.

With small self-excited generators it is important to specify a fault current maintenance kit consisting of a voltage sensitive relay and compounding current transformers (CTs) designed to maintain excitation current when the output voltage collapses owing to a fault.

Differential protection is rarely used on transformers offshore. However, it does have the advantage over restricted earth fault protection in that both primary and secondary windings are protected from both earth and phase-to-phase faults. The relay will only operate for faults appearing in the protection zone between the sets of sensing CTs at the primary and secondary terminals. Under normal load conditions the CT secondary current are equal and no current flows in the relay operating coil. If these currents become unequal owing to a fault current being sensed in the transformer windings, the resulting current energizes the operational coil. The relay contacts close when the ratio of this differential current to the through current exceeds the slope of the relay operating characteristic. This characteristic may be altered by adjusting the turns ratio of the operating and restraint coils.

The bias slope is chosen so that the relay remains insensitive to the spill current produced by through currents flowing to loads or faults external to the transformer. Spill currents occur due to errors that arise in the matching of current transformers and other components in the relay circuit. In addition, a fairly high bias slope is required to prevent maloperation by CT differential currents arising from:

1. Tap changing on transformers giving CT mismatch;
2. Different CT ratios and hence saturation levels giving differential currents under through fault conditions;
3. Magnetizing inrush giving secondary currents in one set of CTs only.

To prevent maloperation by magnetizing inrush, the relay operation is delayed by a selected time delay which allows the initial current peaks to decay to below the relay setting value determined by the percentage bias.

A faulty transformer must be isolated by the circuit breakers on both the primary and secondary sides, wherever the fault is sensed, for two reasons. First, a fault sensed by protection only on the secondary circuit breaker would leave the transformer energized in a dangerous state. Secondly, a fault sensed by protection only on the primary circuit breaker could allow the transformer to be back fed from the secondary circuit, again leaving the transformer energized in a dangerous state.

A scheme where the primary and secondary circuit breaker trip circuits are linked is known as intertripping. Momentary paralleling and intertripping schemes are often combined, and the scheme may have additional facilities, such as those required for the emergency back feeding of a main switchboard air compressor when main generation is not available.

The heating effects due to motor overloads can be simulated to some degree by a bimetal element or an electronic thermal replica. Both the stator windings and the rotor will begin to heat up at the onset of an overload. At ratings below about 20kW, depending on the design and type of machine, the stator windings will tend to suffer thermal damage in a shorter time than the rotor, in which case the motor is termed 'stator critical'. Stator critical motors are easier to protect than rotor critical types since the stator windings are more accessible than the rotor, and can be fitted with temperature sensors if the thermal overload device is felt to be inadequate. If the motor is driven by a variable speed drive, where it is likely to run for some time at low speed, the thermal effects of reduced ventilation must be taken into account in the choice of motor and the setting of the overload. The variable speed drive unit is usually fitted with sophisticated overload protection as a standard feature.

Electronic replica protection relays are normally reserved for the protection of larger machines. However, a drive of only a few tens of kilowatts which has a particularly critical function, or is located at some part of the platform where any form of overheating could not be tolerated. For example, an Ex 'e' motor in a hazardous area, may also require such a device to be fitted.

The supply system depends on a limited number of platform generators which are, in turn, dependent for their fuel, cooling water etc. on the correct operation of the other equipment on the installation. Therefore there is a significant probability that frequency and voltage fluctuations will occur of sufficient magnitude to affect the running of motors. With low-frequency problems, stalling protection will adequately deal with the situation. High-frequency problems should be catered for by overload protection devices.

However, with large reciprocating compressors it may be advisable to fit timed overfrequency and/or underfrequency trips to avoid running the machine for long periods at a frequency close to tolerance limits. Generator under/overfrequency trip settings should be a compromise between taking the most sensitive plant into account, and introducing the likelihood of generator nuisance tripping.

Running at abnormally low system voltage will cause motors to draw large currents and increase heating effects in motor windings and supply cables. A lower system voltage will also increase the likelihood of stalling, as the motor torque is proportional to the square of the voltage at a given speed.

An abnormally high voltage will reduce heating effects in windings and supply cables, and give faster starting times owing to the increased torque. Because of this, it is the practice on some installation to run the system at a few per cent above the nominal voltage. Unfortunately, prospective fault currents increase in proportion to the square of the voltage. Therefore, if the system is to be normally run at a voltage several per cent above nominal, the prospective fault currents should be recalculated to reflect this, and switchgear fault ratings checked accordingly.

The Busbar and Overcurrent Fault

Busbar faults in medium- and low-voltage metal clad indoor switchgear are very rare in onshore installations. However, several such incidents have occurred offshore. In one instance, where the risk of busbar faults was considered too low to warrant the installation of specific protection tripping was initiated by fire detection rather than upstream electrical protection devices.

Water from leaking pipes, or from driving rain or sea spray penetrating module walls, may enter switchgear enclosures, giving rise to explosive faults or serious fires. Arcing faults on the busbars

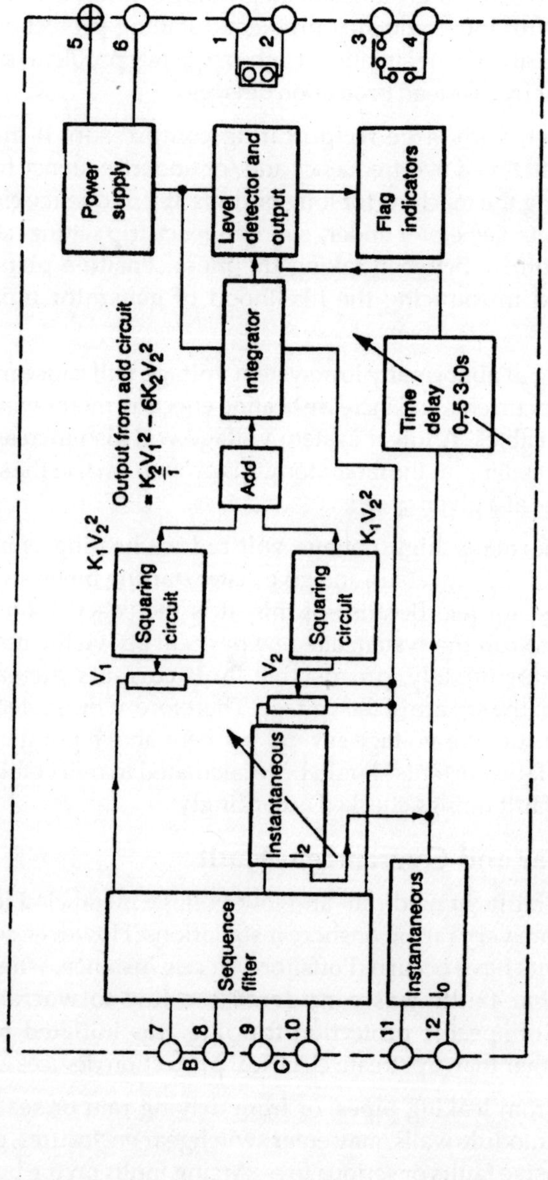

Figure13.1: Schematic of a Typical Electronic Thermal Replica Relay

of low-voltage switchgear will cause extensive damage to the switchboard and may lead to fire in an area of the installation likely to be close to accommodation modules or process control room.

The offshore installation of a full bus zone protection scheme, with check and supervision relays, may only be advisable on the main switchboards of the largest platforms, owing to the weight of all CTs required and the extra space taken up by such a scheme. Nevertheless, it is important that every part of the power system is adequately covered by the protection scheme, and the busbars of switchboards are no exception to this rule.

The overcurrent fault protection relays on the primary circuit of the transformer will provide some protection to the secondary circuit. However, earth fault protection devices on the upstream side will not provide any secondary circuit protection as there is normally no zero sequence path through the transformer.

If the busbars are sectionalized, the bus section switch may be fitted with overcurrent or directional overcurrent protection. Plain overcurrent relays operating on the bus section switches cannot, of course, remove the faulted section from the supply, unless it is being fed via another section. However, if the switchboard is fed by plain feeders or directly by generators with overcurrent protection, such an arrangement can provide reasonable overcurrent protection whilst preserving the supplies on healthy bus sections.

If the switchboard has three sections, directional overcurrent relays, arranged to detect fault currents flowing outwards, may be fitted to the bus section switches. Such an arrangement will preserve the two healthy sections provided the fault is not on the central section. As central section faults relay on upstream protection for

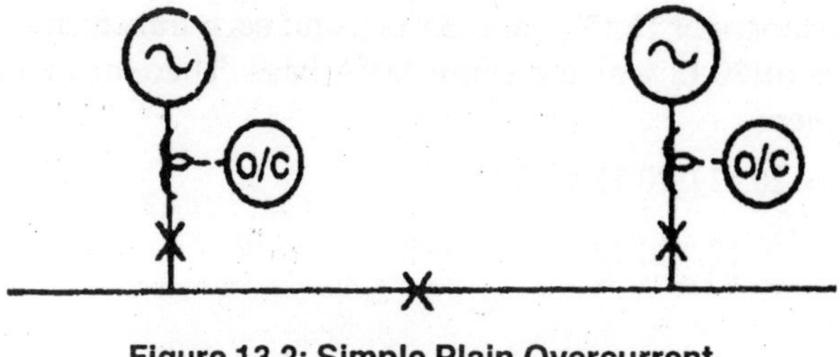

Figure 13.2: Simple Plain Overcurrent Busbar Protection Scheme

clearance, this section should be kept as short and uncomplicated as possible for good reliability. This arrangement could be used for the connection of a third generator, with all outgoing feeders (except third-generator auxiliaries) fed from the end sections.

The Controlled Overcurrent Relays

Some overcurrent relays provide dual inverse characteristics. The less sensitive characteristic gives a longer time interval for clearance of downstream faults, and the more sensitive characteristic remove close-up faults more quickly. Faults close to the generator have less impedance and hence produce greater voltage dips. At a set level of voltage dip, the relay will switch from one characteristic to the other.

In this case we will set the voltage to discriminate between a fault upstream (giving rise to a large voltage dip) and a fault downstream of a transformer (producing a smaller voltage dip). The inset on Figure 13.3 shows the equivalent diagram. The maximum voltage dip occurs when the transformer reactance X_t is a minimum compared with the generator reactance X_g. In this case we will assume this is when one generator is operating and two transformers are paralleled. The voltage at the generator (high-voltage) terminals is given by:

$$V = \frac{X_t}{(X_t + X_g)} - E$$

where,

E is the generated voltage.

The generator reactances from say 0 to 2 seconds can be calculated from the generator decrement curves or from values given by the manufacturer. In this case the generators are identical, each with reactances of $15.1\%_{t=0}$ and $33\%_{t=1}$, and each transformer has a reactance of 20.12% at the same MVA base. Therefore with one transformer:

$V_{t=0} = 20.12/(20.12 + 15.1) = 57\%$

$V_{t=1} = 20.12/(20.12 + 33) = 38\%$

With two transformers in parallel:

$V_{t=0} = 10.06/(10.06 + 15.1) = 40\%$

$V_{t=1} = 10.06/(10.06 + 33) = 23.4\%$

Therefore a setting of 20% should prevent the relay from switching to the more sensitive characteristic on a low-voltage fault.

Two further points may be made. First, this type of relay is known as a voltage controlled overcurrent relay, and should not be confused with the voltage restrained type where the sensitivity is continuously variable over a voltage range. The voltage restrained type is not recommended since sensitivity increases too quickly with dipping voltage so that discrimination with downstream fuses is lost.

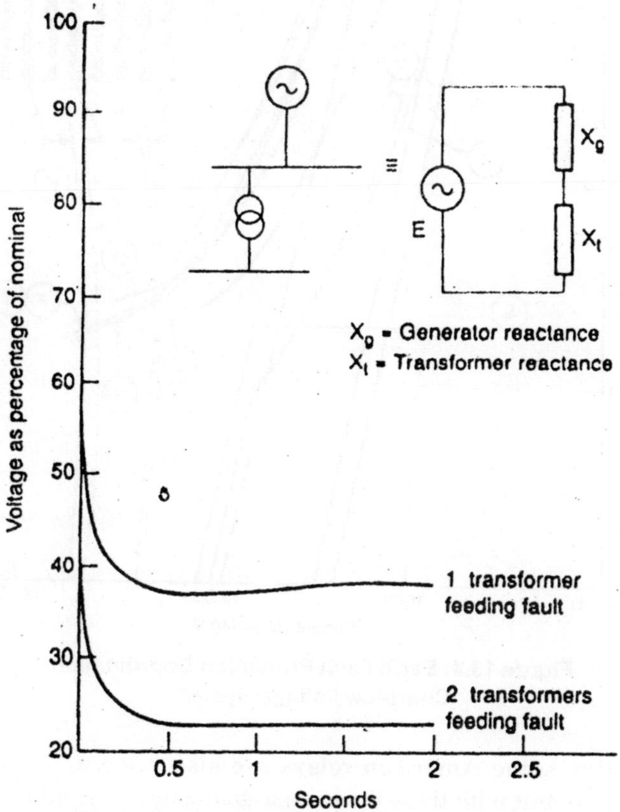

Figure 13.3: Graph of Voltage Downstream of Transformers During System Fault

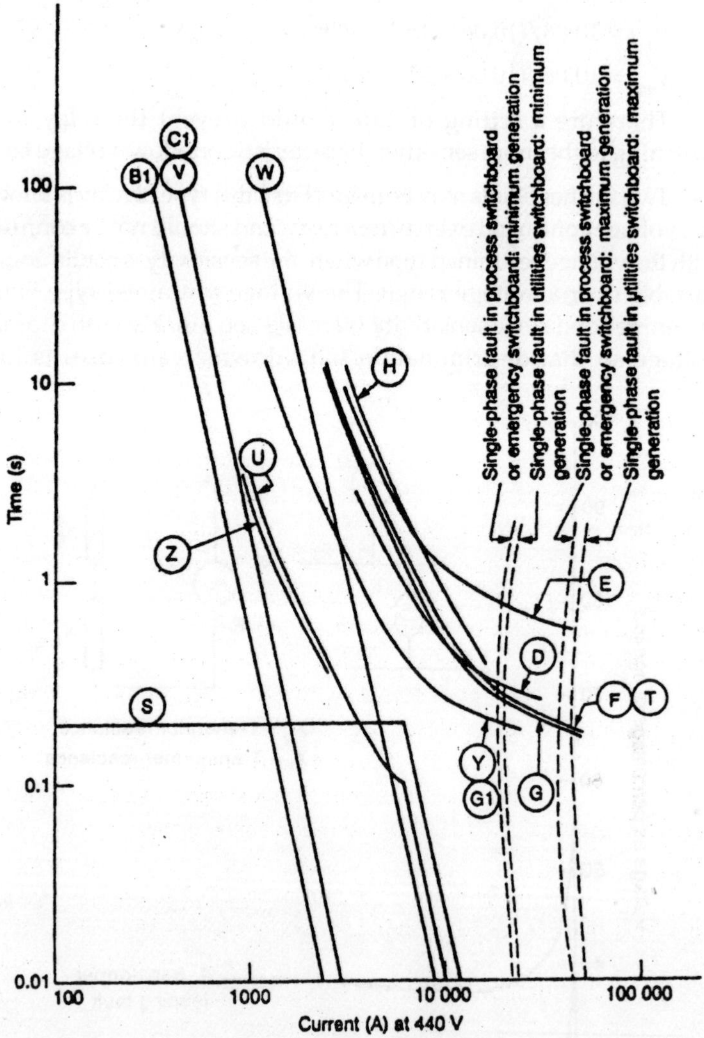

Figure 13.4: Earth Fault Protection Coordination Chart-low Voltage System

Secondly, some American relays are also known as voltage controlled, but with these the inverse characteristic is inhibited at the low levels of voltage dip, leaving only a definite time delay element to clear the fault.

Different Types of Protections of Electrical Equipments

Frame Earth Protection

This type of system has been used successfully offshore, and consists of a frame earth relay and neutral check relay, arranged to monitor earth fault current flows into and out of the switchboard. A simple frame earth system is illustrated in Figure 13.4. Care must be taken, when installing the switchboard into the module, that it is and will remain insulated from the steel of module construction. The generator earthing resistor and the switchboard frame should be earthed at the same point, to avoid series earth connections and to reduce the risk of fault current being limited below the operating setting of the relay by a poor connection due to corrosion of the decking steel etc. This is less likely to be a problem than in an onshore substation using earth electrodes. However, earth connections should be maintained and checked for resistance value on a routine basis, particularly in order to prevent dangerous potentials appearing on the switchboard frame during fault conditions.

An earth or bonding bar interconnects the framework of each cubicle, but is only earthed via the conductor monitored by the frame earth relay. Where the circuit breakers are in trucks, the truck must also have a heavy current earth connection to the bonding bar in order to allow earth fault current flow from the circuit breaker.

To prevent currents induced in cables from causing spurious tripping, the gland plates at the switchboard end should be insulated from the switchboard frame but earthed to a second separate earth bar which is directly earthed to the main earth point.

Unrestricted Earth Fault Protection

On main generator switchboards, either IDMT or definite time earth fault characteristics may be used. The use of definite time is often preferable, as fault current magnitudes are restricted by the generator neutral earthing resistors. Any relay used should be designed so as not to be sensitive to third-harmonic currents, if the generator winding configuration is such that significant third-harmonic currents may circulate.

Low-voltage switchboards are more likely to suffer arcing faults than medium-voltage switchboards. This is because, with increasing voltage, the reduction in fault current due to arc resistance becomes

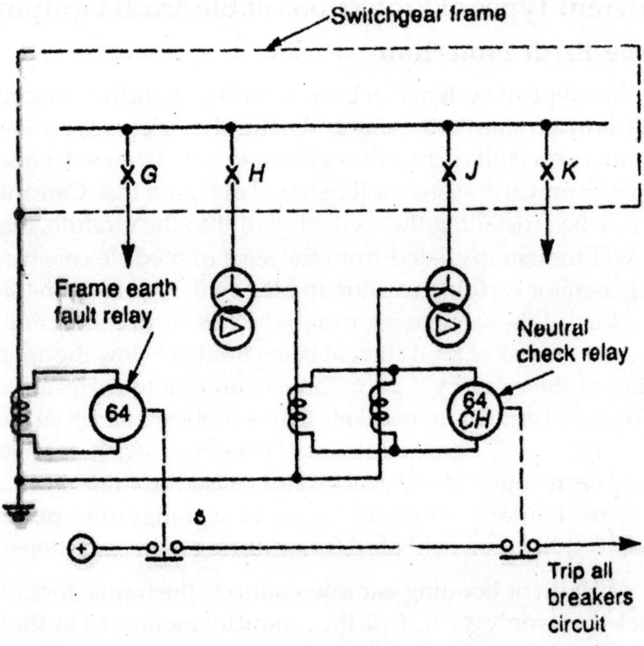

Figure 13.5: Bushbar Frame Earth Protection Scheme

less pronounced as the arc voltage ceases to be a significant proportion of the fault circuit driving voltage. Hence the importance of earth fault protection for low-voltage switchboards is stressed.

Differential Protection

Kirchhoff's first law may be directly applied by monitoring the incoming and outgoing currents at all the switchboard external circuit connections, summing them and detecting the error, should an internal switchboard fault exist. The switchboard may be divided into series of zones, and each zone boundary treated in the same way. The relay settings must be such that the transient flux produced by through fault current will not cause tripping because of imbalance due to unequal burden or saturation of CT's.

By the use of high-impedance relays with series stabilizing resistors, the voltage across the relay during through faults, known as the stability voltage, may be kept below the relay operating voltage. The minimum value of the stabilizing resistor for stability may be calculated as follows:

$$V_S = \text{relay setting current} \times (R_{SR} + R_R)$$

where,

V_S is the stability voltage, R_{SR} is the resistance of the stabilizing resistor and R_R is the resistance of the relay coil.

Low-voltage distribution systems in offshore installations are protected by fuses in the same way as their onshore counterparts. Although the *IEE Wiring Regulations* specifically excludes offshore installations from its scope, the document that it refers to *(IEE Recommendations for the Electrical and Electronic Equipment of Mobile and Fixed Offshore Installations)* is not sufficiently detailed for stand-alone use, and therefore both documents are used for guidance. The methods of calculation given in the *IEE Wiring Regulations* are normally adopted when designing lighting and small power distribution systems, particularly for accommodation modules.

Miniature Circuit Breakers (MCBs)

In most situations, miniature circuit breakers may be used as an alternative to fuses where the fault rating of the MCB is sufficiently high. However, because of the different shape of the MCB tripping characteristic compared with the fuse, it is not advisable to mix fuses with MCBs in the same circuit if discrimination is required. The degree of discrimination between one MCB and another, and between MCBs and fuses, varies with the BS 3871 MCB type, but generally a better discrimination can be obtained using fuses.

Miniature circuit breakers with fault current rating in excess of 16kA are now available and are in common use offshore, where they save weight and reduce the quantity and type of spare fuses stocked. MCBs used offshore must provide positive indication of contact clearance, and must be padlockable when in the open position.

Overcurrent and Earth Fault Protection

Most of the conventional forms of short cable feeder protection may be used offshore. The use of IDMT overcurrent and residually connected earth fault relays may give discrimination problems if a large proportion of the platform load is on the switchboard being fed. This is because the steady state fault current available from the platform generators will be of the order of three to four times the full load current; therefore, if fault clearance times are to be kept short,

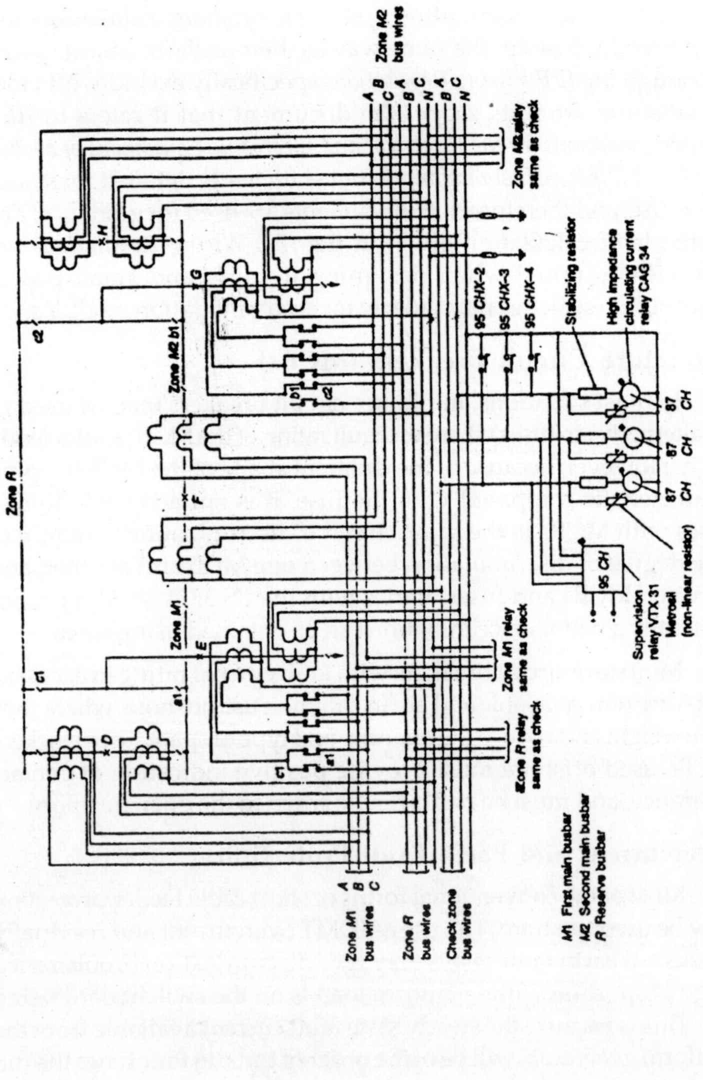

Figure 13.6: Typical Busbar Zone Protection Scheme

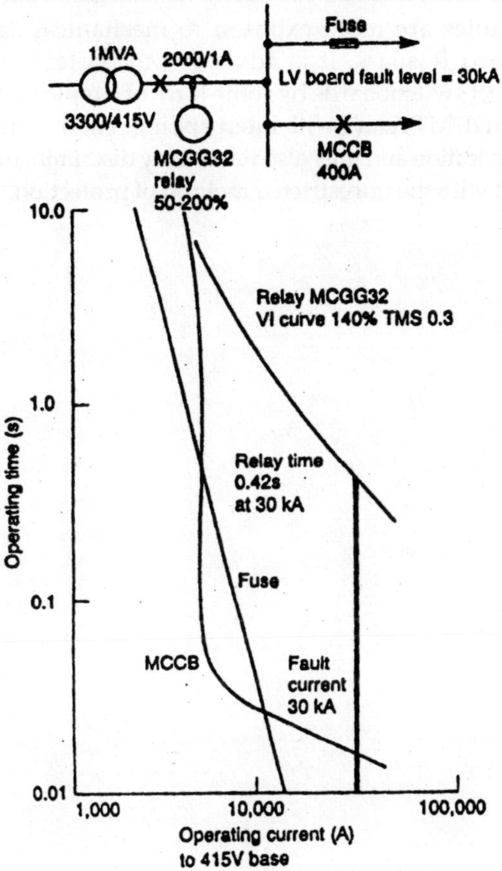

Figure 13.7: Miniature Circuit Breaker and Fuse Characteristic Comparison

generator and feeder relay current settings will need to be close. Voltage controlled relays may be used on generator circuit breakers to overcome this problem, as discussed earlier for generator protection.

An alternative is to use definite time relays. However, on low-voltage systems it is less feasible to use definite time relays, since, if generator and the main switchboard clearance times are to be kept reasonably short, there are likely to be discrimination problems with fuses and MCBs lower down in the system.

As cables are more exposed to mechanical damage than switchboard busbars, it is advisable to protect cables which interconnect switchboards by some form of simple unit protection, rather than IDMT relays with intertripping. This has the advantage of faster operation and may also relieve any discrimination problems associated with the unrestricted method of protection.

Chapter 14

Offshore Cabling Systems

Introduction

The first and most dramatic application of ocean engineering and construction was the bold laying of communication cables across the oceans, undaunted by the depths and in most cases able to span the irregularities of the seafloor by intentionally laying in a loose snake-line pattern as opposed to a straight line. These early cables were not buried. Almost equally remarkable was the development of the ability to locate, grapple for, and pick up severed ends, splice them with waterproof splices, and relay.

In recent years, the need has developed to lay electrical power cables across open water areas. The currently most advanced application is the construction of four cables across the English Channel, trenched in the limestone seafloor. Proposed but not yet built are the power lines across the Strait of Bell Isle, between Newfoundland and Labrador, an irregular rock seafloor gouged and polished by the keels of icebergs. Another deep power cable has been constructed to link the principal Hawaiian Islands.

Key elements in the technological development of subsea cables have been the perfection of armored cable using double-wound helices of steel over the insulated copper or aluminium cable and development of rock-trenching machines capable of cutting a narrow trench up to 1 m deep in hard rock while laying a guide cable in the bottom of the trench. These cutting wheels, fitted with diamond or carbide teeth, must be supported by outrigger sleds or tracked carriages which run on the seafloor. Therefore, the seafloor must have been first cleaned of surficial sediments and debris by one or more of the dredging or jetting schemes discussed in the section of dredging, for example, the use of a trailer suction hopper dredge.

The cutter-trencher, essentially a wheel, cuts a trench 200 to 300 mm wide and meter or so deep. As it moves on, it lays a wire rope line into the bottom of the cut trench. This remains there to later serve as a guide for the power cable. When the trench is complete, the power cable is laid. High-pressure water jets operate from a sled which is guided by the pre-laid wire line. These jets clean out any sediments. The sled device resembles a pipeline trenching sled and may incorporate eductors or airlifts in addition to the jets. The power cable is then fed into the trench, using the previously laid wire line as a guide.

The power cable differs from the pipeline in its much greater flexibility, which enables it to reeled onto a drum of moderate radius and subsequently unreeled without damage. However, as power cables grow in size and armoring, their flexibility decreases, so that they require a laying barge not too dissimilar from a reel barge for pipe laying.

A review of the above procedure will highlight several critical features. First, the seafloor must be either naturally level or artificially leveled to the extent that the trencher can cut a relatively uniform depth. This may require blasting of some outcrops and ledges, using shaped charges or even drilling. The problem with explosive is that the remaining rock is fractured, which can cause the trenching cutter to bind or jam. Mechanical cutting and leveling, with a cuter head suction dredge or ladder dredge is, of course, preferable when water depths permit their use and rock is able to be mechanically excavated.

In many seafloor soils, a heavy plow, such as that described earlier, may prove suitable for direct trenching or for pre-leveling to enable the wheel to work. The methods needed to clean and level the seafloor prior to trenching will vary widely, depending on the character, depth, and hardness of the seafloor surficial materials. A sled device, similar to the large pipeline plows, may be applicable, dragging boulders and loose rock aside. If blasting becomes necessary at discontinuities in profile such as ledges, then closely spaced minimal charges should be used in order to avoid deep fracturing and the resultant large blocks of broken rock.

The power cable crossings of the England Channel consist of four cables, with two being constructed by the British using the methods described above. The seafloor was cleaned by a trailer suction hopper dredge. In some areas, bucket dredges were used to clear the route of boulders.

In recent years, the technology for the design and manufacture of steel-reinforced plastic hoses has advanced significantly so that reliable lines for crude oil and petroleum product service are now available up to at least 16-in diameter. These lines consist of several layers of neoprene, double-wrapped with helixes of steel in a manner resembling that of armored power cables. Being flexible, lines of the size normally used for flow lines can be transported on reels. Large sizes may be faked out on the deck of a barge.

These pipelines have to be led off the barge over a curved cradle, not unlike a small cantilevered stinger. Rollers are used to prevent abrasion. Tension must be applied to control the lowering. Since such lines are usually laid with their end open to the sea, they become progressively flooded. This prevents collapse due to hydrostatic pressure. The net weight of the line may therefore approach that of a steel line. Such flexible lines are usually so short that they are laid in 1 day or less; hence weather conditions may be selected to minimize the required tension. In very deep water or where the tension value exceeds the allowable value for that line, a wire rope line may be married to it to give it support and to take the tension force.

Great care must be used to avoid cutting or abrading such flexible lines. Nylon or other fabric must be used for the slings with which the pipe is handled.

Coflexip has developed an integrated system using a large-diameter wheel trencher on bottom crawlers to enable the simultaneous trenching and laying of the line. Such a system was employed for laying flow lines and cables in the caprock seabed of the Zakum Field, in Abu Dhabi. Coflexip has also developed a method to install its flexible pipe in 450 m of water in the Montanazo Field off Barcelona, Spain. It utilized a very sophisticated positioning system, with a reference array of 30 subsea transponders, and an ROV to monitor pipe touchdown.

Submarine pipelines have been constructed of a wide variety of materials designed to suit the specific environment, operations, and criteria involved. At other places, steel pipelines were addressed, since these constitute the bulk of deep-water pipelines for the transport of oil and gas. Their selection is based both on the need to accommodate high internal pressures in service without leakage and the bending and combined stresses developed during installation.

Other materials widely used for submarine pipelines are polyethylene, fiber-reinforced glass, concrete, and flexible composites of steel and neoprene. Cables for submarine transmission of electric power are another type of flexible pipeline, involving somewhat similar offshore construction techniques.

Fiber-reinforced glass (FRG) has been used for a number of sewer outfalls and seawater intakes, as well as for slurry transport of salt. The material is highly resistant to chemical attack. It can be coated or pigmented to protect it from UV degradation.

It is light in weight, hence even relatively large diameter (2-m) lines can be readily handled in long length, set by a derrick barge, and joined or, alternatively, pulled out from the beach. Because it is of low density, it is usually pulled fully-floated (open-ended). As it is placed, precast concrete saddles are usually lowered over the line to hold it in position under wave action. Because of its light weight, FRG lines are especially hard to lay and hold in place in the surf zone. Weight can temporarily be added internally–for example, chain wrapped with canvas and/or fiber rope so as not to abrade the interior. Even sandbags have been used.

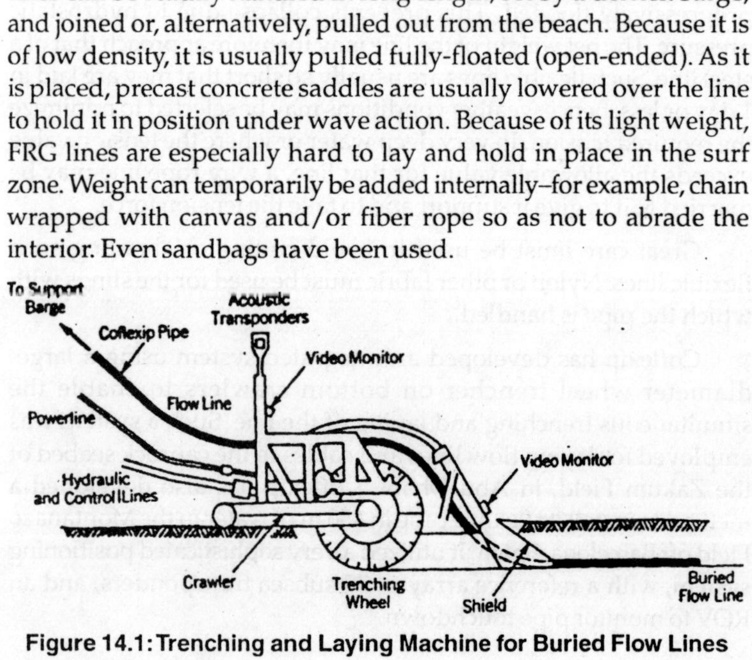

Figure 14.1: Trenching and Laying Machine for Buried Flow Lines

As with polyethylene lines, details of saddle bearings must be carefully developed and accurately constructed to prevent local wear and point bearing. FRG is susceptible to abrasion and to internal fatigue under sustained stress concentrations.

Polyethylene can be used to produce a highly flexible and collapsible line which is resistant to chemical attack and has low friction. It has therefore been utilized for test installations for ocean thermal energy conversion (OTEC) plants, so far involving relatively small diameter (1.5 m and less). This material has a density less than seawater and hence has to be weighted or anchored down to stay properly submerged.

One installation for an OTEC test plant off the island of Hawaii utilized polyethylene pipe made up in long floating strings in quite water and towed to the site, where anchors with short Kevlar lines were used to progressively submerged it to where it floated just above the rough and steeply sloping rock floor. As the weight were attached, the polyethylene line progressively collapsed (buckled) but regained its circular shape as it went underwater, so the buckle traveled along the line as it was laid, without permanent damage.

Reinforced saddles of polyethylene were fitted at regular intervals where the mooring lines were attached to hold the line near the seafloor in an inverted catenary. Stress concentrations at moorings and fittings must be avoided, as polyethylene is subject to internal fatigue under sustained high stress. An advantage of the polyethylene line over a steel line is that no ferrous ions are picked up by seawater; thus it is suitable for use in supplying aquaculture facilities.

Another, large polyethylene line was successfully suspended in a vertical hanging position under an OTEC test facility to a depth of over 1000 m. Later, when this test had been successfully completed, the line has been upended to horizontal and was being towed to a new test site. Reportedly, the heavy end was supported by wire lines; during the tow and under the wave action, these chafed and the concentrated stresses on the polyethylene led to a local failure and loss of the line. (It was later salvaged.)

The Cable Routing Diagram

For the smaller, less complex systems, racking and cable diagrams may be combined. However, in the more complex arrangements, it is more informative to present the information on two or more separate drawings so that the service, route, identification number etc. of each cable can be easily identified.

Separate diagrams may be produced for each service, such as power, instrumentation, communication, fire and gas etc. The disadvantage with such separation is that there will be less likelihood of spotting a clash if one exists, unless an integrated CAD system is being utilized which can identify such clashes. Cable Transit Sections are required in order to determine the size and number of cable transits.

The Cable Schedules

A large offshore installation may have hundreds of kilometres of cables installed, and it is important that, during the design stage, as much information as possible on the identity, route size and type of each cable is retained. This information will obviously be needed for installation, but is also vital should it be necessary to trace, reroute or replace a cable some time later, owing, for example, to fire or mechanical damage. If a particular motor is uprated, it will be necessary to look at the schedule to check weather the existing cable has the capacity for the increased load.

Small project cable schedules may be produced manually using draughting blanks. However, a more practical medium is the computer database, especially during a large project when large numbers of cables are continually being added. For ease of identification, cable should be grouped into service and area. Insertion of extra cables into a schedule would be extremely laborious on a manual schedule but is relatively easy on a database. Other advantages are that duplicate cable numbers cannot exist and that most database programs have powerful sorting and search facilities.

The cable schedule should also contain a list of cable types, giving a full description of each type, a type code that can be referred to in the main schedule, and a total length for each type. Drum and cutting schedules should also be produced in order to minimize cable wastage during installation. The cable schedule must state how the individual cable lengths were derived. For example, do they already contain a margin for drawing inaccuracy, cutting wastage, installation errors etc.?

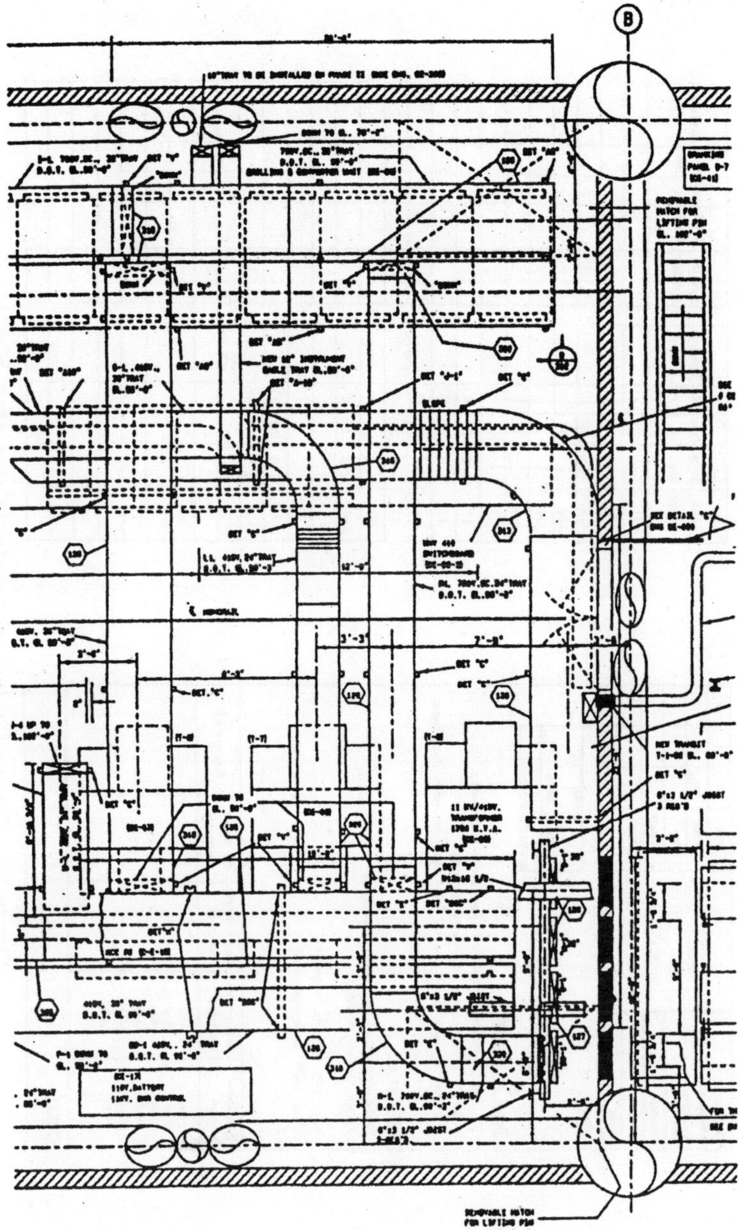

Figure 14.2: Part of a Typical Cable Racking Diagram

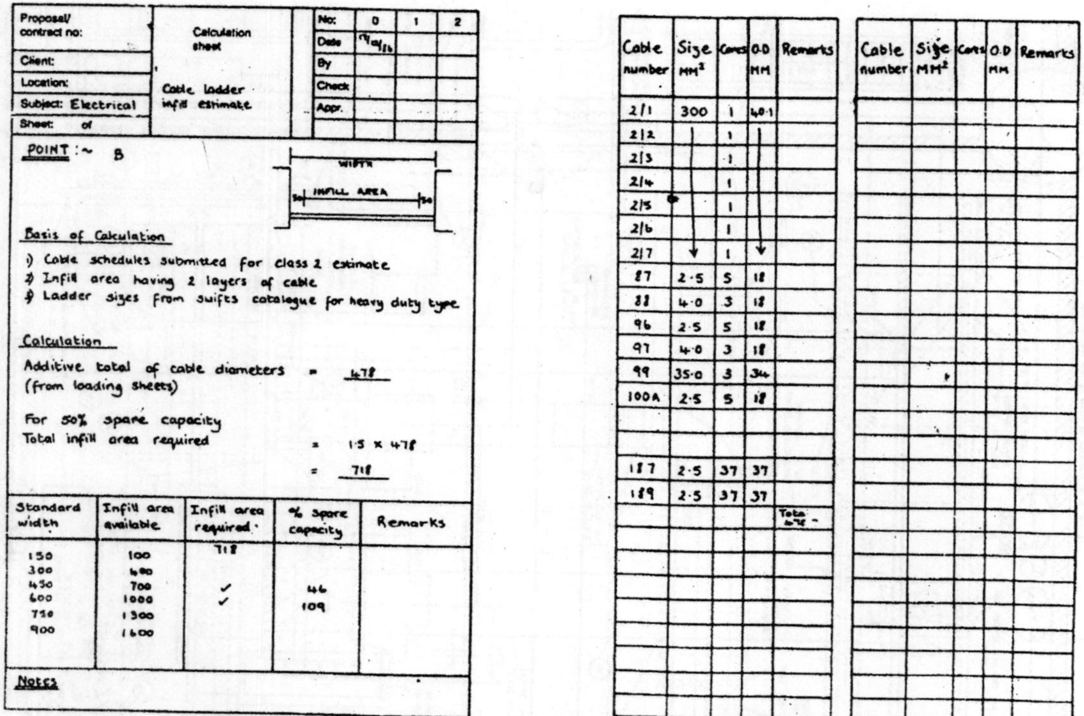

Figure 14.3: Typical Cable Rack Loading Calculation Sheets

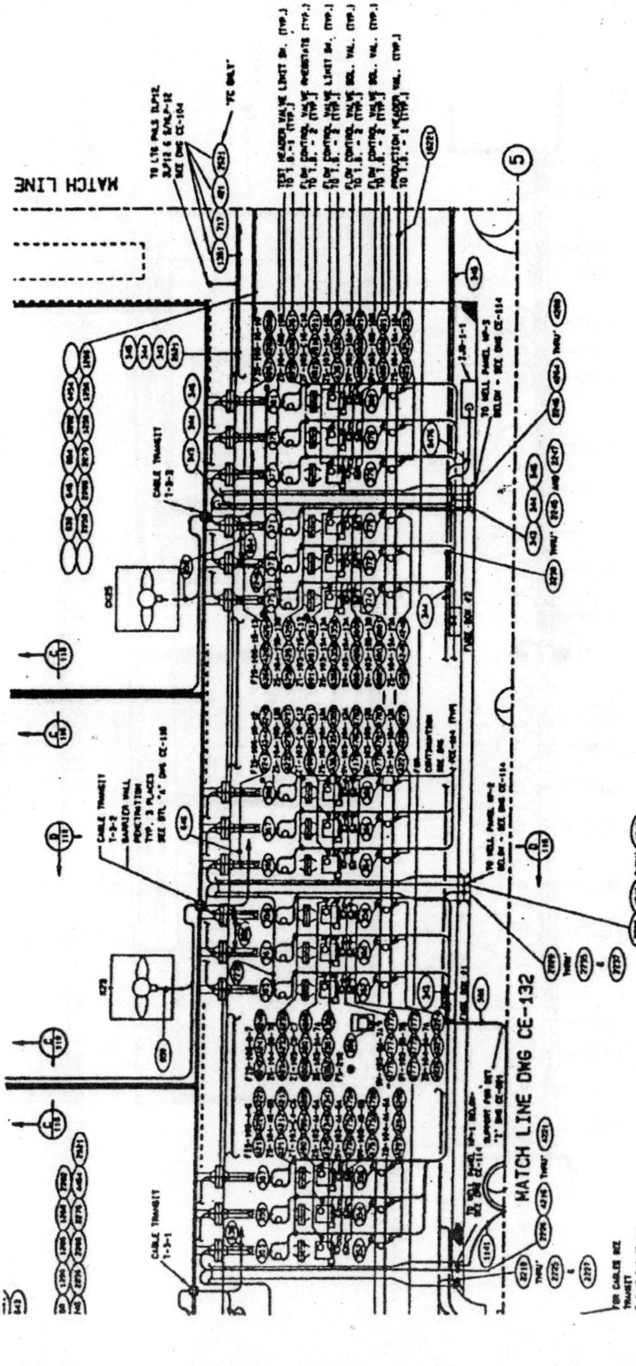

Figure 14.4(a): Parts of a Typical Cable Routing Diagram

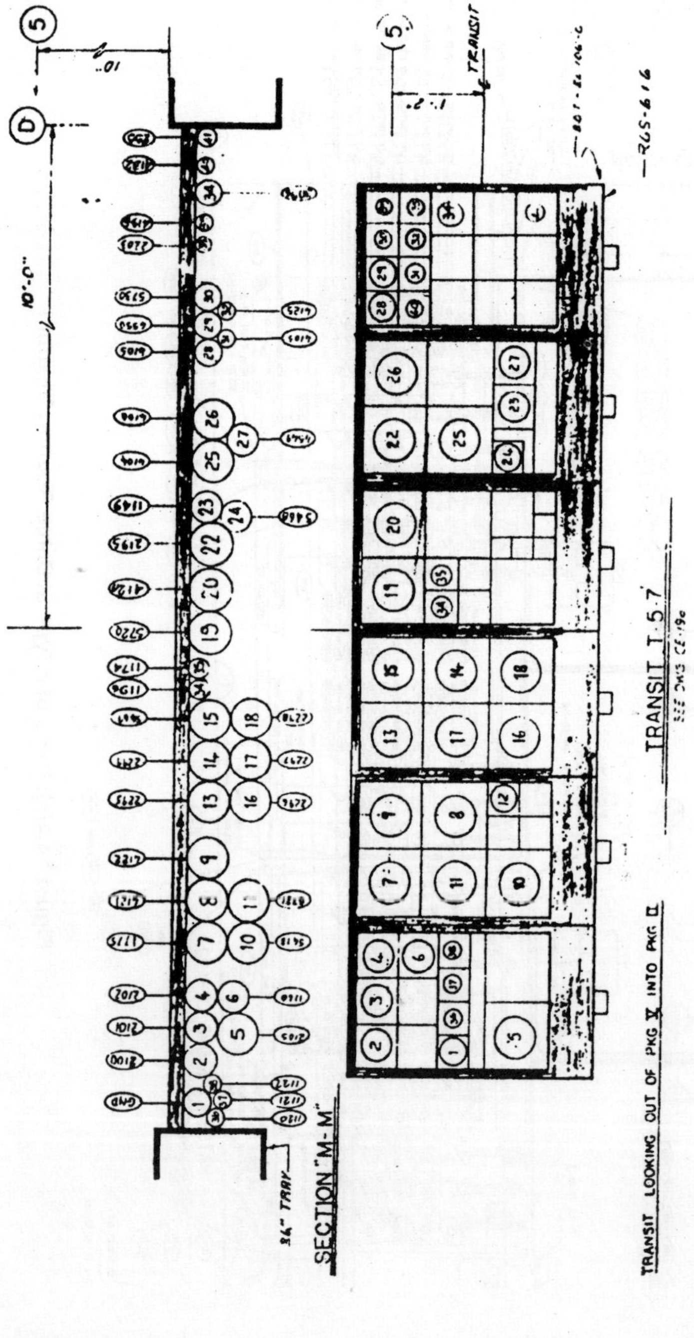

(b)

Figure 14.4(b): Section and Transit Detail

Electrical Equipment Schedule

As an aid to draughting and ensuring that every node in the electrical distribution is catered for in the design, an electrical equipment schedule should be produced. This may be combined with a master equipment list later on in the project. Every item should be allocated a client tag number, and the information presented should include the service, location, manufacturer, type and environmental details such as ingress protection (IP) number and suitability for hazardous areas. A motor schedule will give details of the controls and instrumentation required at the motor, at the starter and at any other point of control. It should also indicate the motor full load current, the fuse size and any special requirements such as earth fault relays or thermistors. The schedule may be split into types of equipments such as motors, junction boxes and luminaries. In some cases it will only be beneficial to produce schedules for particular types of equipment, depending on their populations and whether the same information has been produced on another document.

Again, a computer database is recommended for the production of these schedules.

Installation of Motors and Generators; Cable Support Systems

In most cases, rotating machinery will be installed as a complete skid mounted package. Whether they are installed as a separate items or not, great care must be taken to ensure that the motor and driven machinery are correctly located and aligned after installation. The equipment must be slung or lifted from prescribed lifting positions such as eyebolts where these are available. A normal rule of thumb is that the lifting orientation of the machinery should be identical to its operating orientation , *i.e.* normally lift in the upright position.

During the entire installation procedure, it is important that any tools or materials supplied specifically for the installation of the particular machine should be used. It is not uncommon for such items to be left in the equipment packing case or otherwise overlooked, and for installation to then go ahead using improvised methods which may jeopardize the success of the project and possibly the safety of offshore personnel.

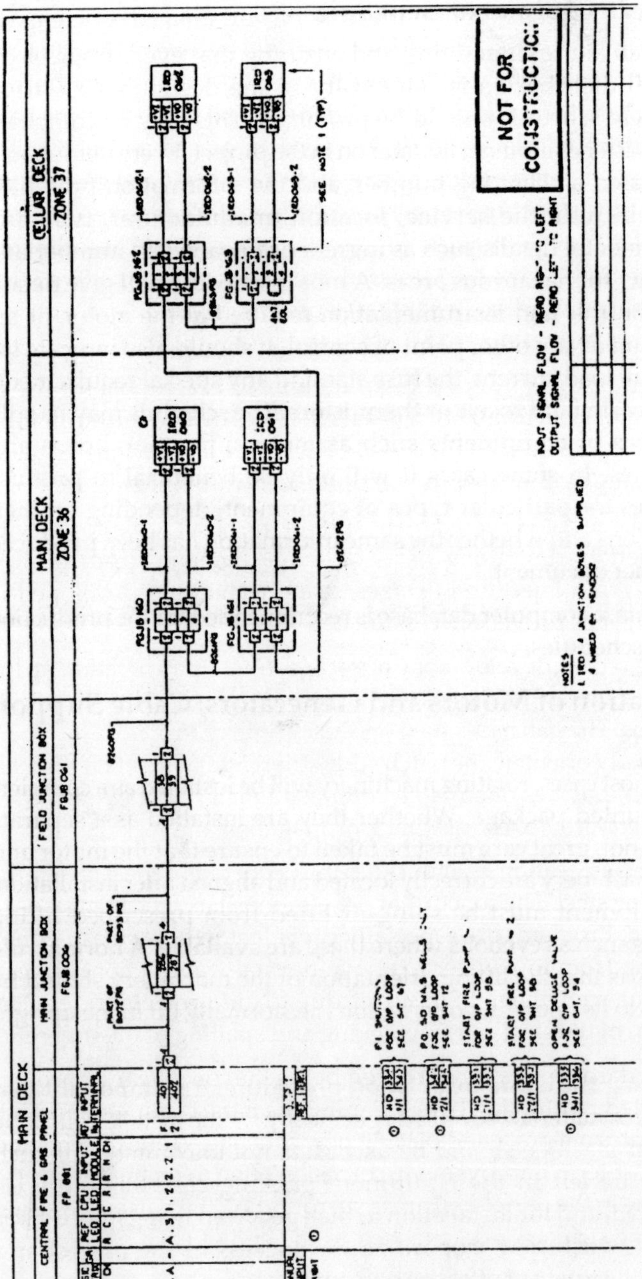

Figure 14.5: Typical Hookup Diagram

Protective films on machined surfaces should be cleaned off with the recommended solvent. This cleaning should take place immediately prior to installation, where plinths are in contact with baseplates etc., and prior to commissioning for such items as motor shafts. Shafts and bearings need to be kept clean and covered whenever work is not proceeding.

CABLE No.	ROUTE FROM EQUIPMENT	ROUTE TO EQUIPMENT	CABLE	CABLE TYPE	No. OF CORES	SIZE mm²	WORKING VOLTS
1	TURBINE GEN. G1 5 MVA	11KV SWITCHGEAR CB-8		A.1.0	3 SINGLE	95	11 KV
2	G2 5 MVA	CB-10					
3	G3 5 MVA	CB-15					
4	11 KV SWITCHGEAR CB-8	11KV/600V TRANSF. T1 2 MVA		A.1.1	1 3/C		
5	CB-18	T2 2 MVA					
6	CB-6	11KV/3.3 KV T3 3.5 MVA					
7	CB-7	T4 3.5 MVA					
8	CB-19	T5 3.5 MVA					
9	CB-5	11KV/415V T6 1750 KVA					
10	CB-20	T7 1750 KVA					
11	CB-21	T8 1750 KVA					
12	11 KV/600V TRANSF. T1	DRILL CONV. INCOMING CUB INC.#1		A.3.4	12 SINGLE E/C	300	600
13	AUX. TRANS.	#1		A.3.3	1-4/C	95	415
14	11 KV/600V TRANSF. T2	#2		A.3.4	12 SINGLE E/C	300	600
15	AUX. TRANS.	#2		A.3.3	1-4/C	25	415
16	11 KV/3.3V TRANSF. T3	3.3 KV SWITCHGEAR CB-32		A.2.1	2-3/C	185	3300
17	T4	CB-34					
18	T5	CB-36					
19	11KV/415V TRANSF. T6	415V CB-41		A.3.4	17 SINGLE E/C	300	415
20	T7	CB-45					
21	T8	CB-47					
22	DEISEL GEN. G6	CB-43			11 SINGLE E/C	150	
23	G7	CB-49					
24	415 VSWGR CB-40 CUB. C	M.C.C. No. 4 YS S3 TIER 'E' UPPER			13 SINGLE E/C		
25	CB-42 CUB. C	M.C.C. No. 4 YS S1 TIER 'E' LOWER					
26	M.C.C. - 1LX1 CB-56	M.C.C. No. 5 COMPT. E1					
27	M.C.C. - 2LA1 CB-57	M.C.C. No. 6 COMPT. A1					
28	TURBINE GENERATOR G1	NEUT EARTHING RESISTOR BOX NER 1		A.1.0	3 SINGLE	95	11 KV
29	G2	2					
30	G2	3					
31	NEUTRAL EARTHING RESISTOR #1	EARTHING SYSTEM		A.3.4	3-1/C		
32	NEUTRAL EARTHING RESISTOR #2	EARTHING SYSTEM			3-1/C		
33	NEUTRAL EARTHING RESISTOR #3	EARTHING SYSTEM			3-1/C		
34	11KV/600V TRANSF. 'T1'	NER 'T1'		A.3.4	1-1/C	300	
35	11KV/600V TRANSF. 'T2'	NER 'T2'		A.3.4	1-1/C	300	
36	NER 'T1'	EARTHING SYSTEM		D.I.F.	2-1/C	95	
37	NER 'T2'	EARTHING SYSTEM		D.I.F	2-1/C	95	
38							
39							

Figure 14.6: Part of a Typical Sheet from a Cable Schedule

Contd...

GRADE	ROUTE FEET	DUTY	EQUIP. No.	HAWKE GLAND CABLE	CABLE ROUTED VIA TRAY	REMARKS
11 KV	375	POWER	C-J-01	D	B7 B5 B2 B3 B B1	RUN CABLES IN 1 TREFOIL GROUP
	350		C-J-02	D	B7 B5 B2 B3 B B1	
	280		C-J-03	D	A7 A5 A6 A3 A A1	
	70		T1	F	LOCAL PKG I	
	50		T2			
	85		T3			
	65		T4			
	65		T5			
	75		T6			
	55		T7			
	40		T8			
1000	50			C2		RUN CABLES IN 4 TREFOIL GROUPS @ 3" BETWEEN GROUPS
	50			C		RUN CABLES IN 4 TREFOIL GROUPS @ 3" BETWEEN GROUPS
	65			C2		
	65			C		
3.3 KV	70			F		3" BETWEEN GROUPS
	100			F		
	95			F		5 TREFOIL GROUPS @ 3" BETWEEN GROUPS (2 CABLES NEUT CONN)
1000	85			C2		
	60			C2		
	60			C2		3 TREFOIL GROUPS @ 3" BETWEEN GROUPS (2 CABLES NEUT CONN)
	40		C-J-06	C		
	40		C-J-07			2 TREFOIL GROUPS @ 3" BETWEEN GROUPS (2 CABLES NEUT CONN)
	300				A7 A5 A6 A3 A A1	
	325				B7 B5 B2 B3 B B1	
	60				LOCAL PKG I	
	65					
11 KV	18	EARTHING		D	B7-1 PKG VII	NEUTRAL CONNECTION
	18			D	B7-2	
	18			D	A7-1	
1000	16			C	LOCAL	EARTHING
	11			C		
	15			C		
	40			C2	LOCAL PKG I	
	40			C2		
	50			C		
	50			C		

Figure 14.6–Contd...

Shims supplied with the equipment should be used wherever possible for leveling. Shims should be of similar area and shape as the machined surfaces of the equipment footings for which they are used, and the maximum thickness of shim should always be used so that the minimum number of shims are positioned under each footing. Where separate bearing pedestals are supplied, these should be positioned, leveled and lined up prior to the installation of other machinery components. White metal bearings should be checked to ensure adequate bedding using engineer's blue or some other suitable method of indication.

Coupling faces should be truly parallel and level. This will need to be checked with each shaft rotated to several different positions. Coupling bolts or flexible connections should be properly fitted without damage and all nuts tightened and locked before running the machinery.

When large machines, and particularly synchronous machines, are installed, the manufacturer's instructions concerning insulation must be carefully followed to prevent circulating currents flowing in the machine frame. Where the machine frame is installed in sections, conductive bonding between each section will need to be established, if necessary by separate bonding conductors. This is particularly important when the motor is cited in a hazardous area, since otherwise sparking may occur between sections.

The larger, heavier machines should be provided with permanent runway beams and horizontal and vertical screw jacks to facilitate alignment. Alternatively, suitable jacking points can be provided for temporary jacks. The use of block and tackle arrangements or tirfors is time consuming and likely to increase the risk of injury to installers.

Before commissioning, machines should be cleaned; if they are of the open type, dust and dirt should be blown out. Commutators and slip rings will need to be checked for deterioration during transit and storage, and cleaned if necessary. Carbon brushes will need to be checked for freedom of movement, and springs checked for correct pressure. If machines have been stored for a long period, the bearing grease should be inspected; if deterioration has occurred, the bearings should be thoroughly washed, dried and regreased using methods recommended by the particular bearing manufacturer. Where oil lubricating systems are employed, bearing oil rings, flow switches, pressure switches, pumps etc. must be checked to ensure they are fully operative before any running of the machine is attempted.

Machine shafts will require to be rotated by hand or barring gear to ascertain that no foreign body is either inside the machine or between the external fan and its protecting cowl. Checks should also be made to ensure that ventilating air ducts are clean and clear of obstruction. Where safety guards are provided with the machinery, these should be fitted prior to any rotation of the equipment under power. The guard should be constructed so as to facilitate removal. The location of both the equipment and the guards on the equipment should be such that access for inspection and maintenance is practicable and safe. Both guards and couplings between motors and machinery should be removable without requiring removal of the motor or the driven equipment. Where guards are over belts and pulleys, a check should be made that the belt does not slap the guard at any speed. This is particularly important with variable speed drives.

Before any rotating equipment is commissioned offshore, the following precautions should be taken:

All necessary tools and ancillary equipment should be conveniently to hand but not so placed as to pose a safety hazard.

Notices relating to fire fighting procedures and treatment of electric shock and burns should be displayed in switchrooms and control rooms.

All normal working and warning notices for the equipment should be clearly visible.

Now let us try to understand about cable support systems. In the design phase, a great deal of care should have been taken to ensure that the correct sizes, configuration and routes have been shown on the cable and racking arrangement drawings. If route clashes with pipework or overloading of supports are to be avoided, it is vital to install both support systems and cable strictly according to design drawings. However, there is always the possibility that a design error has been made. Thus the installer should carefully check the locations as shown on the drawings to determine if any conflict exists between the new cable route and any other equipment, steelwork, piping, ducting etc., and whether, when installed, the way will cause a hazard to personnel, obstruct accessways or prevent the installation or removal of equipment.

On minor routes, not detailed on design drawings, where it is necessary to run cable support systems, it is important that separate racks or trays are used for the following categories in order to prevent electromagnetic interference between cables:

Medium-voltage AC (above 1kV) with associated control cables;

Low-voltage AC (240V to 1000 V) with associated control cables; DC power cables, associated control cables and 110 V DC/254 V AC instrument cables;

Instrument, signal and alarm cables; telephone, communications, and fire and gas cables.

Where practicable a minimum segregation distance of 1 meter should be maintained. Crossovers should be kept to a minimum and as near perpendicular as possible.

Some vital circuits associated with fire and gas, ESD or other safety related systems may have supply and/or signal cables duplicated. Such cables should be run on separate routes so that a single fire cannot destroy both cables.

Cables used on offshore installation will require the following attributes:

They should have stranded copper conductors for smaller cross-section and better flexibility.

They should be voltage and fault rated for the system in which they are operating.

They should be rated for normal maximum current flows of the circuit without exceeding the maximum conductor temperature or the temperature class limit if passing through a hazardous area.

If they are involved in a fire, smoke and acid fume emission should be low. The insulation compound used should have an oxygen index of at least 30.

They should normally be armoured for mechanical protection. Braided or wire armouring may be used; the braided type is slightly more flexible but is often more difficult to gland.

These criteria cannot be rigidly applied, however, since special cables will be needed for instrumentation, diving umbilicals, downhole pumps etc.

The cables designed for the most onerous duty are those for electric fire pump supplies. These fire survival cables are designed to continue to supply current after an hour at a temperature of 1000°C, followed by housing down with high-pressure water jets whilst being hit with hammer. The difficulty lies in finding a suitable support system that will survive the same treatment!

The most common cable type in use is the EPR/CSP type similar to that found in merchant shipping, which meets the high oxygen index and fire retardant requirements of BS 6883 and IEC 93-3.

Cables must be seized to allow for circuit current and maximum voltage drop as with any onshore system. AC cables can be seized using a chart generated by spreadsheet, as shown in Figure 14.7. Care will still need to be exercised to ensure that close excess protection is still applied and that cable fault ratings are adequate.

Although, the IEE Wiring Regulations (sixteenth edition 1991) do not apply to offshore installation, they should be used for guidance in the sizing of conductors, unless more specific information is made available in a future revision of the IEE Recommendations for the Electrical and Electronic Equipment of Mobile and Fixed Offshore Installations (first edition 1983).

During the installation of cables following factors must be remembered:

Cables should be routed in such a way as to facilitate maintenance and the installation of additional cables with minimum need for expensive scaffolding. For example, a cable route located under a deck, so that there is no immediate access and the sea is directly below it, is not recommended. The cost of a small cabling modification in such a location would be overshadowed by the enormous scaffolding bill.

Cable route should avoid known fire where this is possible. Cables to main and standby machinery should be run on separate routes.

Cables should be laid parallel on cable ladders and tray in a neat and orderly fashion.

Where heavy three-phase currents are carried, cables are usually single-core. Single-core cables of the same conductor cross-

CABLE SELECTION FOR 415 VOLT MOTORS
STARTING CURRENT = 6#FLA

Frequency = 60Hz
CABLE SWA/EPR/CSP

KW	FLA AMP	CABLE X-SECTION	CABLE AMP	DISTANCE METRES	VOLT DROP PER METRE	DERATE FACTOR	POWER FACTOR	SIN@	RESISTANCE PER METRE	REACTANCE PER METRE	VOLTAGE KILOVOLTS	reactance @ 50Hz
5.5	9.002	1.500	22.00	75.084	1.105	0.409	0.850	0.527	0.0138	0.000164	0.415	0.000136
5.5	8.897	2.500	33.00	136.454	0.608	0.270	0.860	0.510	0.0076	0.000151	0.415	0.000125
7.5	12.133	1.500	22.00	55.079	1.507	0.551	0.860	0.510	0.0138	0.000164	0.415	0.000136
7.5	12.133	2.500	33.00	100.066	0.829	0.368	0.860	0.510	0.0076	0.000151	0.415	0.000125
11.0	17.794	2.500	33.00	68.227	1.217	0.539	0.860	0.510	0.0076	0.000151	0.415	0.000125
11.0	17.794	4.000	44.00	108.986	0.762	0.404	0.860	0.510	0.0047	0.000149	0.415	0.000124
15.0	24.265	2.500	33.00	50.033	1.659	0.735	0.860	0.510	0.0076	0.000151	0.415	0.000125
15.0	24.265	4.000	44.00	79.923	1.039	0.551	0.860	0.510	0.0047	0.000149	0.415	0.000124
18.5	29.927	6.000	58.00	97.145	0.854	0.516	0.860	0.510	0.0031	0.000142	0.415	0.000118
15.0	24.265	10.000	77.00	199.031	0.417	0.315	0.860	0.510	0.0018	0.000140	0.415	0.000116
22.0	35.180	6.000	58.00	81.787	1.015	0.607	0.870	0.493	0.0031	0.000142	0.415	0.000118
22.0	35.180	10.000	77.00	135.967	0.610	0.457	0.870	0.493	0.0018	0.000140	0.415	0.000116
30.0	47.973	10.000	77.00	99.709	0.832	0.623	0.870	0.493	0.0018	0.000140	0.415	0.000116
30.0	47.973	16.000	103.00	155.021	0.535	0.466	0.870	0.493	0.0012	0.000131	0.415	0.000109
37.0	59.166	16.000	103.00	125.693	0.660	0.574	0.870	0.493	0.0012	0.000131	0.415	0.000109
37.0	59.166	25.000	135.00	192.905	0.430	0.438	0.870	0.493	0.0007	0.000124	0.415	0.000103
45.0	71.141	16.000	103.00	103.644	0.801	0.691	0.880	0.475	0.0012	0.000131	0.415	0.000109
45.0	71.141	25.000	135.00	159.272	0.521	0.527	0.880	0.475	0.0007	0.000124	0.415	0.000103
55.0	86.950	25.000	135.00	130.313	0.637	0.644	0.880	0.475	0.0007	0.000124	0.415	0.000103
55.0	86.950	35.000	165.00	175.520	0.473	0.527	0.880	0.475	0.0005	0.000122	0.415	0.000101
75.0	118.569	35.000	165.00	128.715	0.645	0.719	0.880	0.475	0.0005	0.000122	0.415	0.000101
75.0	118.569	50.000	200.00	167.849	0.494	0.593	0.880	0.475	0.0004	0.000120	0.415	0.0001
90.0	142.282	50.000	200.00	139.874	0.593	0.711	0.880	0.475	0.0004	0.000120	0.415	0.0001
90.0	142.282	70.000	250.00	191.506	0.433	0.569	0.880	0.475	0.0003	0.000117	0.415	0.000097
110.0	173.901	70.000	250.00	156.686	0.530	0.696	0.880	0.475	0.0003	0.000117	0.415	0.000097
110.0	173.901	95.000	300.00	198.604	0.418	0.580	0.880	0.475	0.0002	0.000114	0.415	0.000095
132.0	208.681	95.000	300.00	165.503	0.502	0.696	0.880	0.475	0.0002	0.000114	0.415	0.000095
132.0	208.681	120.000	360.00	202.774	0.409	0.580	0.880	0.475	0.0002	0.000112	0.415	0.000093
160.0	252.946	150.000	400.00	192.407	0.431	0.632	0.880	0.475	0.0001	0.000112	0.415	0.000093
160.0	252.946	185.000	460.00	223.580	0.371	0.550	0.880	0.475	0.0001	0.000112	0.415	0.000093
200.0	316.183	185.000	460.00	178.864	0.464	0.687	0.880	0.475	0.0001	0.000112	0.415	0.000093
200.0	316.183	240.000	550.00	211.013	0.393	0.575	0.880	0.475	0.0001	0.000111	0.415	0.000092
250.0	395.229	300.000	630.00	191.553	0.433	0.627	0.880	0.475	0.0001	0.000110	0.415	0.000091

Figure 14.7: Spreadsheet Generated Cable Sizing Chart

section usually have a high fault rating th their three-phase equivalents, and are easier to install, havin 'ower weight per metre and a lower bending radius. To avoi y currents being induced in local steelwork, such cables must be run in a trefoil configuration. In some situations the cables must be run in a flat profile because of space limitations, in which case a balanced arrangement must be adopted which still avoids the promotion of eddy currents. Support or other steelwork must pass between phases, as the steel will heat up owing to hysteresis loss. Where two or more trefoil cable groups run on the same route, they should be at the same horizontal level with a clear space between the groups of at least one cable diameter.

Cable bends should not be lighter than the minimum bending radius specified by the manufacturer. Drip loops should be provided at external cable terminations. These will also be useful if the cable has to be reterminated at some later date. In any case, straining of cables at cable glands should be avoided, and cables should be perpendicular to gland plates for a minimum of 100 mm before entering the gland. Expansion loops will also be required across the expansion gaps necessary between module walls and support steelwork.

Some cable insulation materials tend to become brittle with lower temperatures. This is particularly the case with PVC compounds. Under low-temperature conditions, care needs to be taken when installing most cable types. It is desirable that cable with PVC insulation should not be installed directly at ambient temperatures of 5°C and below. In such conditions, the cable should be stored for at least the previous 24 hours at a temperature of 20°C or more.

Cables should be clamped or cleated to rack or tray with suitable ties. The usual arrangement is to use nylon 12 ties (not nylon 66 as it becomes brittle in the offshore environment), supplemented at suitable intervals (depending on the weight) by EVA coated stainless steel banding or cable cleats of the correct diameter. Ties should be non-magnetic for single-core cables cleated in flat formation.

Where trefoil cleats are used for single-core cables, stainless steel banding should also be fitted for fault current bracing at 1 metre intervals, with a band always located close to each cleat.

The positive and negative cables forming a DC circuit should be run side by side as far as possible to reduce magnetic effects.

Spare cores in multicore cables should be coiled back for future use, not cut off at the gland.

Index

G

H

I

L

M

R

S